Nano-Optics and Nanophotonics

Editor-in-Chief

Motoichi Ohtsu, Tokyo, Japan

Editorial Board

The Springer Series in Nano-Optics and Nanophotonics provides an expanding selection of research monographs in the area of nano-optics and nanophotonics, science- and technology-based on optical interactions of matter in the nanoscale and related topics of contemporary interest. With this broad coverage of topics, the series is of use to all research scientists, engineers and graduate students who need up-to-date reference books. The editors encourage prospective authors to correspond with them in advance of submitting a manuscript. Submission of manuscripts should be made to the editor-in-chief, one of the editors or to Springer.

More information about this series at http://www.springer.com/series/8765

Motoichi Ohtsu · Takashi Yatsui
Editors

Progress in Nanophotonics 3

 Springer

Editors
Motoichi Ohtsu
Takashi Yatsui
Graduate School of Engineering
International Center for Nano Electron
  and Photon Technology
The University of Tokyo
Tokyo
Japan

ISSN 2192-1970   ISSN 2192-1989   (electronic)
Nano-Optics and Nanophotonics
ISBN 978-3-319-11601-3   ISBN 978-3-319-11602-0   (eBook)
DOI 10.1007/978-3-319-11602-0

Library of Congress Control Number: 2014953515

Springer Cham Heidelberg New York Dordrecht London

Springer is part of Springer Science+Business Media (www.springer.com)

Preface to *Progress in Nanophotonics*

As the first example, recent advances in photonic systems demand drastic increases in the degree of integration of photonic devices for large-capacity, ultrahigh-speed signal transmission and information processing. Device size has to be scaled down to nanometric dimensions to meet this requirement, which will become even more strict in the future. As the second example, photonic fabrication systems demand drastic decreases in the size of the fabricated patterns for assembling ultra-large-scale integrated circuits. These requirements cannot be met even if the sizes of the materials are decreased by advanced methods based on nanotechnology. It is essential to decrease the size of the electromagnetic field used as a carrier for signal transmission, processing and fabrication. Such a decrease in the size of the electromagnetic field beyond the diffraction limit of the propagating field can be realized in optical near fields. Nanophotonics, a novel optical technology that utilizes the optical near field, was proposed by M. Ohtsu (the editor of this monograph series) in 1993 in order to meet these requirements. However, it should be noted that the true nature of nanophotonics involves not only its ability to meet the above requirements. It is also its ability to realize qualitative innovations in photonic devices, fabrication techniques, energy conversion and information processing systems by utilizing novel functions and phenomena made possible by optical near-field interactions, which are otherwise impossible as long as conventional propagating light is used. Based on interdisciplinary studies on condensed matter physics, optical science, and quantum field theory, nanomaterials and optical energy transfer in the nanometric regime have been extensively studied in the last two decades. Through these studies, novel theories on optical near fields have been developed, and a variety of novel phenomena have been found. The results of this basic research have been applied to develop nanometer-sized photonic devices, nanometer resolution fabrication, highly efficient energy conversion, and novel information processing, resulting in qualitative innovations. Further advancement in these areas is expected to establish novel optical sciences in the nanometric space, which can be applied to further progress in nanophotonics in order to support the sustainable development of people's lives all over the world. This unique monograph series entitled Progress in Nanophotonics in the Springer Series in Nano-optics and Nanophotonics is introduced to review the results of advanced studies in the field of nanophotonics and covers the most recent topics of theoretical and experimental interest in relevant fields, such as classical and quantum optical sciences, nanometer-sized condensed matter physics, devices, fabrication techniques, energy conversion, information processing, architectures, and algorithms. Each chapter is written by leading scientists in the relevant field. Thus, this monograph series will provide high-quality scientific and technical information to scientists, engineers, and students who are and will be engaged in nanophotonics research. As compared with the previous monograph series entitled Progress in Nano-Electro-Optics (edited by M. Ohtsu, published in the Springer Series in Optical Science), this monograph series deals not only with optical science on the nanometer scale, but also its applications to technology. I am grateful to Dr. C. Ascheron of Springer-Verlag for his guidance and suggestions throughout the preparation of this monograph series.

Tokyo   Motoichi Ohtsu

Preface to Volume 3

This volume contains five review articles focusing on various but mutually related topics in nanophotonics written by the world's leading scientists. The first article describes light-emitting diodes and lasers made of indirect transition-type silicon bulk crystals in which the light emission principle is based on dressed-photon–phonons. A novel phenomenon called photon breeding is also reviewed. The second article is devoted to describing theoretical studies on optoelectronic properties of molecular condensates for organic solar cells and light-emitting devices. It also discusses the physical origins underlying organic photovoltaics and light emissions for rationalizing the experimental results. The third article describes the basics of topological light beams together with their important properties for laser spectroscopy. It also presents experimental results on nonlinear four-wave mixing and pump probe reflection spectroscopy, by which unique properties associated with topologically ordered electrons in materials are investigated. The fourth article is devoted to review spatially localized modes emerging in nonlinear discrete dynamic systems. It also describes the essential mechanism for the existence of intrinsic localized modes and their basic properties. The last article describes theoretical methods to explore the dynamics of nanoparticles by the light-induced force of tailored light fields under thermal fluctuations. These methods will pioneer a new research field based on fluctuation mediated nano-optomechanics.

This volume is published with the support of Prof. Yatsui of the University of Tokyo, an Associate Editor. I hope this volume will be a valuable resource for readers and future specialists in nanophotonics.

Tokyo

Motoichi Ohtsu

Contents

Contributors

Yusuke Doi Department of Adaptive Machine Systems, Graduate School of Engineering, Osaka University, Suita, Osaka, Japan

Takuya Iida Graduate School of Science, Osaka Prefecture University, Osaka, Japan

Syoji Ito Graduate School of Engineering Science, Osaka University, Toyonaka, Osaka, Japan

Tadashi Kawazoe Graduate School of Engineering, The University of Tokyo, Tokyo, Japan

Masayuki Kimura Department of Electric Engineering, Kyoto University, Nishikyo-ku, Kyoto, Japan

Chie Kojima Graduate School of Engineering, Osaka Prefecture University, Osaka, Japan

Ryuji Morita Department of Applied Physics, Hokkaido University, Sapporo, Japan; Core Research for Evolutionary Science and Technology (CREST), Japan Science and Technology Agency (JST), Kawaguchi, Japan

Motoichi Ohtsu Graduate School of Engineering, International Center for Nano Electron and Photon Technology, The University of Tokyo, Tokyo, Japan

Hiroyuki Tamura WPI-Advanced Institute for Material Research (AIMR), Tohoku University, Sendai, Japan

Yasunori Toda Department of Applied Physics, Hokkaido University, Sapporo, Japan; Core Research for Evolutionary Science and Technology (CREST), Japan Science and Technology Agency (JST), Kawaguchi, Japan

Shiho Tokonami Nanoscience and Nanotechnology Research Center, Osaka Prefecture University, Osaka, Japan

Kazuyuki Yoshimura NTT Communication Science Laboratories, NTT Corporation, Sorakugun, Kyoto, Japan

Chapter 1
Silicon Light Emitting Diodes and Lasers Using Dressed Photons

Motoichi Ohtsu and Tadashi Kawazoe

Abstract This chapter reviews light emitting diodes and lasers made of indirect-transition-type silicon bulk crystals in which the light emission principle is based on dressed photons. After presenting physical pictures of dressed photons and dressed-photon–phonons, the principle of light emission by using dressed-photon–phonons is reviewed. A novel phenomenon named photon breeding is also reviewed. Next, the fabrication and operation of light emitting diodes and lasers are described, in which the role of coherent phonons in these devices is discussed. Finally, light emitting diodes using other relevant crystals are described, and other relevant devices are also reviewed.

1.1 Introduction

This article reviews how to use an indirect-transition-type semiconductor to construct a light emitting diode (LED) and a laser, which has not been possible by conventional methods employed in materials science and technology. The wavelength of the light emitted from a conventional LED is governed by the bandgap energy E_g of the semiconductor material used. Although there is a Stokes wavelength shift [1], its magnitude is negligibly small. Therefore, the value of E_g must be adjusted for the desired light emission wavelength, and this has been achieved by exploring novel semiconductor materials. For this purpose, direct transition-type semiconductors have been used for conventional LEDs. Among these semiconductors, InGaAsP has been used for optical fiber communication systems because the wavelength of the emitted light is $1.00–1.70\,\mu m$ ($E_g = 0.73–1.24\,eV$) [2, 3]. When fabricating a highly efficient infrared LED or laser using InGaAsP, it is necessary to use a double

M. Ohtsu (✉)
Graduate School of Engineering, International Center for Nano Electron and Photon Technology, The University of Tokyo, 2-11-16 Yayoi, Bunkyo-ku, Tokyo 113-8656, Japan
e-mail: ohtsu@ee.t.u-tokyo.ac.jp

T. Kawazoe
Graduate School of Engineering, The University of Tokyo, 2-11-16 Yayoi, Bunkyo-ku, Tokyo 113-8656, Japan
e-mail: kawazoe@ee.t.u.-tokyo.ac.jp

© Springer International Publishing Switzerland 2015
M. Ohtsu and T. Yatsui (eds.), *Progress in Nanophotonics 3*,
Nano-Optics and Nanophotonics, DOI 10.1007/978-3-319-11602-0_1

heterostructure composed of an InGaAsP active layer and an InP carrier confinement layer. However, there are some problems with this approach, including the complexity of the structure and the high toxicity of As [4]. In addition, In is a rare metal. On the other hand, composite semiconductors that emit visible light, such as AlGaInP and InGaN, have extremely low emission efficiencies around the wavelength of 550 nm (=2.25 eV) [5], which is called the green gap problem. Although this efficiency has been increasing recently due to improvements in dopant materials and fabrication methods, there are still several technical problems because highly toxic or rare materials are required, which increases the cost of fabrication.

In order to solve the problems described above, several methods using silicon (Si) have been recently proposed. Although Si has been popularly used for electronic devices, there is a long-held belief in materials science and technology that Si is not suitable for use in LEDs and lasers because it is an indirect-transition-type semiconductor, and thus, its light emitting efficiency is very low. The reason for this is that electrons have to transition from the conduction band to the valence band to spontaneously emit light by electron–hole recombination. However, in the case of an indirect-transition-type semiconductor, the momentum (the wavenumber) of the electron at the bottom of the conduction band and that of the hole at the top of the valence band are different from each other. Therefore, for electron–hole recombination, a phonon is required in the process to satisfy the momentum conservation law. In other words, electron–phonon interaction is required. However, the probability of this interaction is low, resulting in a low interband transition probability.

In order to solve this problem, for example, porous Si [6], a super-lattice structure of Si and SiO_2 [7, 8], and Si nanoprecipitates in SiO_2 [9] have been used to emit visible light. To emit infrared light, Er-doped Si [10] and Si–Ge [11] have been employed. In these examples, the emission efficiency is very low since Si still works as an indirect-transition-type semiconductor in these materials.

In contrast to these examples, as explained in this article, the use of dressed photons (DPs) and dressed-photon–phonons (DPPs) can realize highly efficient LEDs and lasers even when using Si bulk crystal.

1.2 Dressed Photons and Dressed-Photon–Phonons

In the conventional quantum theory of light, the concept of a photon was established by quantizing the electromagnetic field of light propagating through a macroscopic free space whose size is larger than the wavelength of light. A photon corresponds to an electromagnetic mode in a virtual cavity defined in free space. Since a photon is massless, it is difficult to express its wave function in a coordinate representation in order to draw a picture of the photon as a spatially localized point particle, like an electron. Thus, interactions between photons and electrons in a nanometric space must be carefully investigated.

To describe a light field in a nanometric space, the energy transfer between two nanomaterials and detection of the transferred energy are essential. They are formulated by assuming that the nanomaterials are arranged in close proximity to each other and illuminated by propagating light. Although the separation between the two nanomaterials is much shorter than the optical wavelength, it is sufficiently long to prevent electron tunneling. As a result, the energy is transferred not by a tunneled electron but by some sort of optical interactions between the two nanomaterials.

A serious problem, however, is that a virtual cavity cannot be defined in a sub-wavelength-sized nanometric space, unlike the conventional quantum theory of light. In order to solve this problem, an infinite number of electromagnetic modes, with an infinite number of frequencies, polarization states, and energies, must be assumed. In parallel with this assumption, infinite numbers of energy levels must also be assumed for the electrons. As a result of these assumptions, the dressed photon (DP) is defined as a virtual photon that dresses the material energy, i.e., the energy of the electron [12, 13]. The interaction between the two nanomaterials can be represented by energy transfer due to the annihilation of a DP from the first nanomaterial and its creation on the second nanomaterial.

Because of the infinite number of electromagnetic modes described above, the DP field is modulated temporally and spatially. The temporal modulation feature is represented by an infinite number of modulation sidebands, i.e., an infinite series of photon energies. Furthermore, since an actual nanometric system (composed of nanomaterials and DPs) is always surrounded by a macroscopic system (composed of macroscopic materials and electromagnetic fields), energy transfer between the nanometric and macroscopic systems has to be considered when analyzing the inter-action between the nanomaterials in the nanometric system. As a result, it is found that the magnitude of the transferred energy is represented by the Yukawa function. This function represents the spatial modulation feature of the DPs. This analysis also elucidates an intrinsic feature of DPs, namely, the size-dependent resonance; that is to say, the efficiency of the energy transfer between nanomaterials depends on the sizes of the nanomaterials that are interacting. It should be noted that this resonance is unrelated to diffraction, which governs the conventional wave-optical phenomena. Furthermore, since the DP is localized in nanometric space, the long-wavelength approximation, which is valid for conventional light–matter interactions, is not valid for DP-mediated interactions. As a result, an electric dipole-forbidden transition turns out to be allowed in the case of the DP-mediated interactions.

In actual materials, such as semiconductors, the contribution of the crystal lattice also needs to be included in the theoretical model of the DP. By doing so, it has been found that the DP interacts with phonons, which are quanta of normal modes of the crystal lattice vibration. Furthermore, in a nanomaterial, it is possible to gen-erate multi-mode coherent phonons via a DP–phonon interaction. As a result of this interaction, a novel quasi-particle is generated. This quasi-particle is called a dressed-photon–phonon (DPP), which is a DP dressing the energy of the multi-mode coherent phonon. Furthermore, as a result of spatial modulation, the DPP field is localized at the impurity sites in the crystal lattice of the nanomaterial with a spatial extent as short as the size of the impurity site. It is also localized at the end of the nanomaterial with a spatial extent as short as the size of this end.

The DPP energy can be transferred to the adjacent nanomaterial, where it induces a novel light–matter interaction. Here, since translational symmetry is broken due to the finite size of the nanomaterial, the momentum (wave-number) of the quasi-particle has a large uncertainty and is non-conserved, as was the case of the DP itself. Also, the DPP field is temporally and spatially modulated. As a result of the temporal modulation, the DPP gains an infinite number of modulation sidebands. As a dual relation of this modulation, the electron dresses the energies of the photon and phonon, which means that the energy of the electron in the nanomaterial is also modulated.

When analyzing the conventional light scattering phenomenon in a macroscopic material, it has been sufficient to study one phonon. In contrast, the coherent phonon described above is composed of an infinite number of phonons. This coherent phonon assists in exciting the electron in the adjacent nanomaterial instead of merely increasing the material temperature, which enables DPP-assisted excitation of the electrons. Therefore, for analyzing the optical excitation of the nanomaterial, it is essential to represent the relevant quantum state of the nanomaterial by the direct product of the quantum states of the electron and the coherent phonon. This relevant quantum state means that an infinite number of energy levels has to be considered in the energy bandgap between the valence and conduction bands of an electron in the case of a semiconductor. This quasi-continuous energy distribution originates from the modulation of the eigen-energy of the electron as a result of the coupling between the DP and the coherent phonon. This DPP-assisted excitation and de-excitation can be exploited in the fabrication and operation of LEDs, lasers, and other devices using Si bulk crystals, which are reviewed in the following sections.

1.3 Principles of Light Emission

For spontaneous emission of light by DPPs, a single-step or a two-step de-excitation of an electron takes place depending on whether the emitted photon energy is higher or lower than the energy difference between the excited and ground states of the electron, which is E_g in the case of a semiconductor material. Stimulated emission also takes places in the same way.

1.3.1 Single-Step De-excitation

In Fig. 1.1a, b, since the electron strongly couples with the photon and phonon, the energy state is expressed as the direct product of the ket vectors of the electronic state and the phonon state. For example, $|E_g; el\rangle \otimes |E_{ex}; phonon\rangle$ includes the ground state of the electron and the excited state of the phonon. The spontaneous emission of a DPP, as well as propagating light (Fig. 1.1a), is the result of the radiative transition from the excited state $|E_{ex}; el\rangle \otimes |E_{ex}; phonon\rangle$ to the ground state $|E_g; el\rangle \otimes |E_{ex'}; phonon\rangle$. Here, in the case of a semiconductor, the excited and ground states of the electron, ($|E_{ex}; el\rangle$ and $|E_g; el\rangle$), correspond to the states in the conduction and valence bands, respectively. After the transition, the phonon in the excited state relaxes to the thermal equilibrium state ($|E_{thermal}; phonon\rangle$) determined by the crystal lattice temperature, and finally the electron and phonon transition to the state $|E_g; el\rangle \otimes |E_{thermal}; phonon\rangle$.

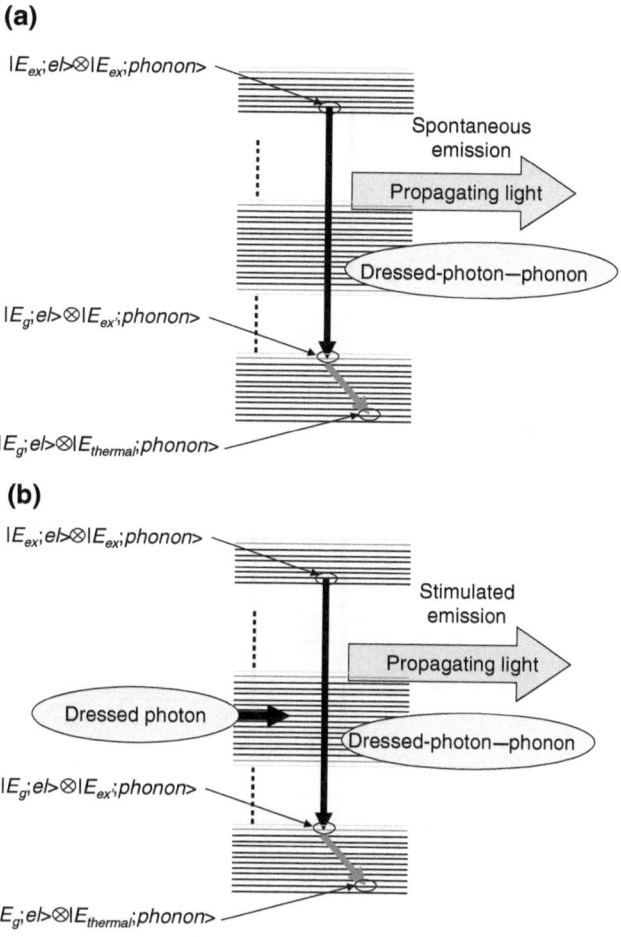

Fig. 1.1 Single step optical emission processes via dressed-photon–phonon. **a** Spontaneous emission. **b** Stimulated emission

The stimulated emission process (Fig. 1.1b) is explained as follows: When an electron in the excited state is irradiated with a DP, a transition from the initial state $|E_{ex}; el\rangle \otimes |E_{ex}; phonon\rangle$ to the ground state $|E_g; el\rangle \otimes |E_{ex'}; phonon\rangle$ takes place, resulting in light emission. Like the spontaneous emission, the phonon then relaxes to a thermal equilibrium state determined by the crystal lattice temperature and finally the electron and phonon transition to the state $|E_g; el\rangle \otimes |E_{thermal}; phonon\rangle$.

1.3.2 Two-Step De-excitation

Figure 1.2a schematically illustrates the two-step de-excitation for spontaneous emission.

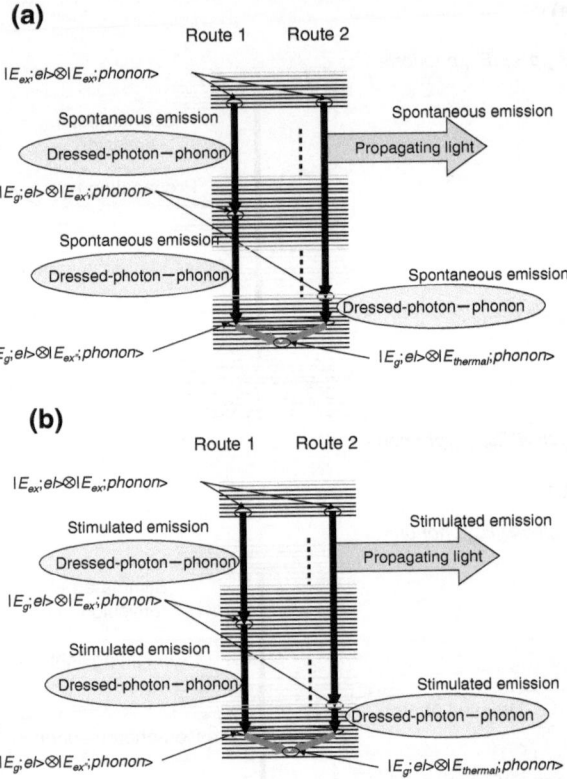

Fig. 1.2 Two-step optical emission processes via dressed-photon–phonon. **a** Spontaneous emission.
b Stimulated emission

1.3.2.1 First Step

The initial state is expressed by the direct product of the excited state of the electron
and the excited state of the phonon, $(|E_{ex}; el\rangle \otimes |E_{ex}; phonon\rangle)$. De-excitation takes
place from this initial state to the ground state $|E_g; el\rangle$ of the electron. Since this
de-excitation is an electric dipole-allowed transition, it generates not only a DPP
but also propagating light. As a result, the system reaches the intermediate state
$|E_g; el\rangle \otimes |E_{ex'}; phonon\rangle$. Here, the excited state $|E_{ex'}; phonon\rangle$ of the phonon after
DPP emission (route 1 in Fig. 1.2a) has a much higher eigenenergy than that of
the thermal equilibrium state $|E_{thermal} : phonon\rangle$. This is because the DP couples
with the phonon, resulting in phonon excitation. On the other hand, the excited state
$|E_{ex'}; phonon\rangle$ of the phonon after propagating light emission (route 2 in Fig. 1.2a) has
an eigenenergy as low as that of $|E_{thermal} : phonon\rangle$. This is because the propagating
light does not couple with the phonon.

1.3.2.2 Second Step

This step is an electric dipole-forbidden transition because the electron stays in the ground state. Thus, only the DPP is generated by this emission process. As a result, the phonon is de-excited to the lower excited state $|E_{ex''}; phonon\rangle$, and the system is expressed as $|E_g; el\rangle \otimes |E_{ex''}; phonon\rangle$. After this transition, the phonon promptly relaxes to the thermal equilibrium state, and thus, the final state is expressed as $|E_g; el\rangle \otimes |E_{thermal}; phonon\rangle$.

The de-excitation for the stimulated emission is explained by Fig. 1.2b, which is similar to Fig. 1.2a. The only difference is that the DP is incident on the electron in the excited state, triggering stimulated emission.

1.4 Visible Light Emitting Diodes

This section reviews the fabrication and operation of a Si bulk p–n homojunction-structured LED [14]. Spontaneous emission takes places based on the principle described in Sect. 1.3.1 because the emitted photon energy is higher than Eg of Si.

Since the DPP has a large uncertainty in its momentum due to its spatially modulated nature (refer to Sect. 1.2), it can provide momentum for electron de-excitation, i.e., for the recombination of an electron and hole. Although the lowest point of the conduction band (X-point) and the highest point of the valence band (Γ-point) correspond to different momenta, an electron in the conduction band efficiently relaxes to the ground state and emits a photon thanks to the assistance of the phonon in the DPP. Furthermore, a radiative transition from a high-energy excited electron can also easily occur via the DPP. For example, due to the existence of the DPP at a high energy (e.g., level a in Fig. 1.3), an excited-state electron nearby can quickly couple with the coherent phonon and then directly relax to the ground state; thus, a radiative relaxation shown by the red downward arrows occurs, resulting in emission with a photon energy higher than E_g. Recall that, without the DPP, an electron in an excited state at high energy quickly transitions to the lowest point in the conduction band

Fig. 1.3 Energy band structure of Si. *Blue horizontal lines* represent the phonon-coupled electronic states. *Red* and *blue downward arrows* show the radiative relaxation

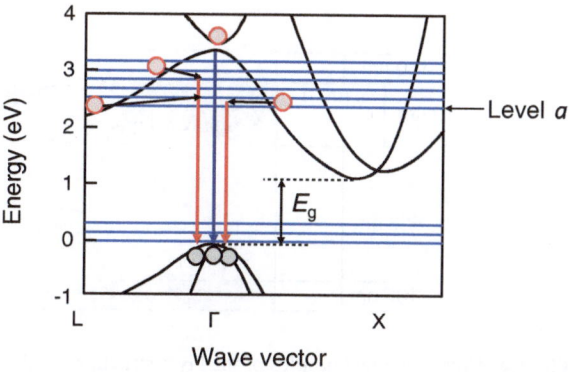

due to fast intraband relaxation; therefore, the probability of an interband transition from a high-energy excited state is extremely low in conventional methods.

In order to realize a visible Si-LED, the single-step spontaneous and stimulated emission processes described in Sect. 1.3.1 were used two times. The first is for fabrication of the device, more specifically, for self-organization of the spatial distribution of the dopant to form a distribution that is suitable for the emission of high-energy photons. The second is for the operation of the device, to obtain spontaneously emitted light. These are reviewed in the following sections.

1.4.1 Device Fabrication

The fabrication of the device can be divided into two steps. The first step is to prepare a Si p–n homojunction structure having a modifiable dopant distribution. The second step is to modify the shape of the dopant domains through DPP-assisted annealing.

In the first step, an epitaxial layer of phosphorus (P) was deposited on an n-type Si crystal with low arsenic (As) concentration. Subsequently, this Si crystal was doped with boron (B) by ion implantation, with seven different accelerating energies, namely, 30, 70, 130, 215, 330, 480, and 700 keV, to form dopant domains with a dopant concentration of 10^{19} cm^{-3}. This doping formed a p-type region in the Si, and as a result, a p–n homojunction structure was constructed. The B distribution in the p–n homojunction was spatially inhomogeneous due to the high accelerating energy and high doping concentration; this was to increase the probability of producing a B distribution favorable for generating DPPs. The crystal was then diced. Then, an ITO film was deposited on the surface of the p-type layer, and a Cr/Al film was deposited on the back surface of the n-type Si for use as electrodes. The cross-sectional profile of the layer structure of the device is shown in Fig. 1.4.

In the second step, DPP-assisted annealing was performed by causing a forward-bias current to flow through the device while irradiating the p-type side of the device with a laser beam. The forward-bias current density used for the annealing was 1.44 A/cm^2. The optical power density of the laser beam was 3.33 W/cm^2, which was sufficiently high for inducing DPP-assisted annealing. The photon energy $h\nu_{anneal}$ of

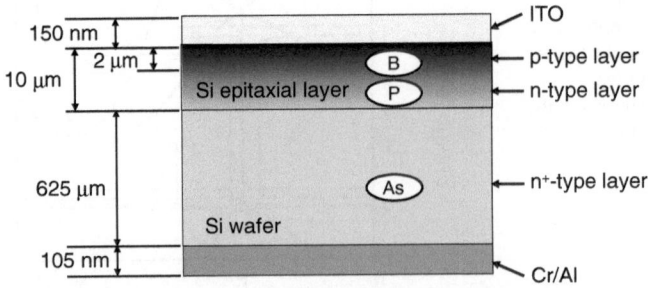

Fig. 1.4 Cross-sectional profiles of the layer structures of the device

the laser light was 3.1 eV (400 nm wavelength), which was 2.5-times higher than E_g of Si (1.12 eV, or 1.11 μm wavelength) [15], and thus the light was absorbed in the Si bulk crystal by a direct transition. The forward-bias voltage was set to be higher than $h\nu_{anneal}/e$, where e represents the electron charge.

It was pointed out in Sect. 1.2 that the DPP is localized at the impurity sites in the nanomaterial with a spatial extent as short as the size of the impurity site. This suggests the possibility of generating and localizing DPPs at the B impurity sites by light irradiation, depending on the spatial distribution and concentration of B. During the DPP-assisted annealing process, B diffuses, and its spatial distribution in the p–n homojunction is continuously modified in a self-organized manner until reaching the desired shape.

The mechanism of this DPP-assisted annealing is explained as follows, by considering regions where DPPs are hardly generated and regions where DPPs are easily generated:

(1) Regions where DPPs are hardly generated: If the device is irradiated with light satisfying $h\nu_{anneal} > E_g$, as shown in Fig. 1.5a, when an electron in the valence band absorbs a photon, it is simultaneously scattered by a phonon (indicated by the upward blue arrow and the green wavy arrows, respectively). As a result, the electron is excited to the conduction band and then immediately relaxes to the bottom of the conduction band. After this relaxation, the electron cannot radiatively relax because of the different momenta between the bottom of the conduction band and the top of the valence band. Therefore, it eventually relaxes non-radiatively, as indicated by the red wavy arrow. This non-radiative relaxation generates heat in the Si crystal, causing the B to diffuse, and thus changing the spatial distribution of the B concentration.

(2) Regions where DPPs are easily generated: When the B diffuses as in (1), the B concentration changes to a spatial distribution suitable for generating DPPs with a fairly high probability. Here, in the case where DPPs are generated at the surfaces of the B domains, as shown in Fig. 1.5b, an electron injected by the forward current is de-excited via the DPP energy level (indicated by horizontal

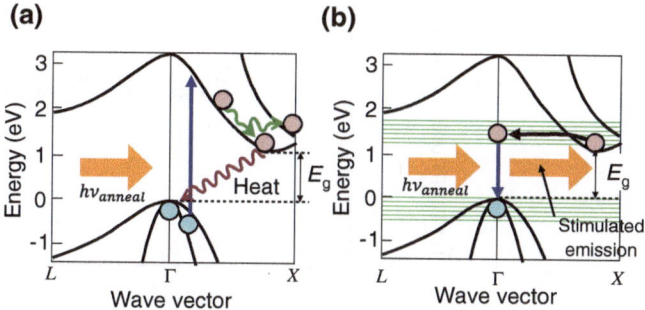

Fig. 1.5 Schematic explanation of the DPP-assisted annealing. **a** Regions where DPPs are hardly generated. **b** Regions where DPPs are easily generated

green parallel lines in this figure), producing a photon generated by stimulated emission. When the electron number densities of occupation in the excited state (the conduction band) and the ground state (the valence band), denoted by n_{ex} and n_g, satisfy the Bernard–Duraffourg inversion condition ($n_{ex} > n_g$) [16], the number of photons created by stimulated emission exceeds the number of photons annihilated by absorption. Since the photons generated by stimulated emission are radiated outside the device, part of the light energy that the device absorbs is dissipated outside the device in the form of propagating light energy, and therefore, the thermal diffusion rate of the B becomes smaller than in (1).

(3) Due to the difference in the thermal diffusion rates between (1) and (2), after the B has diffused throughout the entire device in a self-organized manner as annealing proceeds, it reaches an equilibrium state in which its spatial distribution has been modified, and the annealing process is thus completed. Since region (2) is in a state where stimulated emission with photon energy $h\nu_{anneal}$ is easily generated via the DPPs, and since the stimulated emission probability is proportional to the spontaneous emission probability [17], the DPP-assisted annealed device should become an LED that exhibits spontaneous emission with the photon energy $h\nu_{anneal}$ of the irradiation light. In other words, the irradiation light during the DPP-assisted annealing serves as a "breeder" that generates light with the same photon energy $h\nu_{anneal}$ in the LED; that is, a novel phenomenon that we call photon breeding takes place in this LED.

After 1 h of annealing, fabrication of the Si-LED was completed. To confirm the stimulated emission, Fig. 1.6 shows the measured change in surface temperature difference between the irradiated area and the non-irradiated area. The Si crystal was continuously irradiated with light, and current injection was started after 7 min. At the beginning, when only the laser beam was radiated, the temperature increased to 75 °C, and the temperature difference dramatically increased due

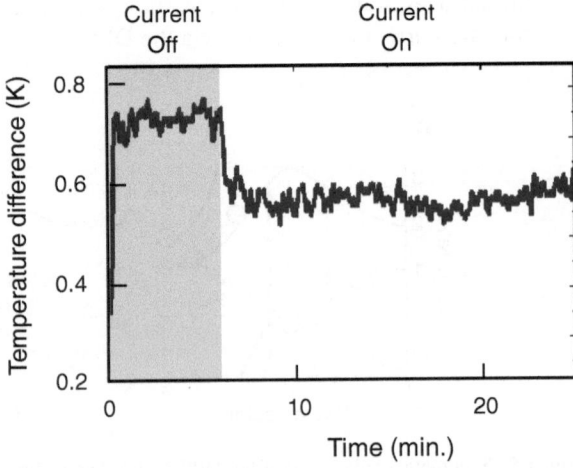

Fig. 1.6 The measured change in surface temperature difference between the irradiated area and the non-irradiated area

to the heat provided by light absorption. After a few minutes, the heat gradually diffused into the whole device, and the temperature difference reached a stable value. This agreed with our hypothesis of how the annealing process advances with laser irradiation, which was described in (1) and (2). Next, after 7 min, a forward-bias voltage was applied to the device, while continuing to radiate the laser beam. This led to an obvious decrease in the temperature difference. This decrease was a result of stimulated emission in the area where DPPs were generated; that is, the irradiated light was not converted into heat but induced light emission due to the stimulated emission process. In other words, this decrease in temperature difference confirmed the occurrence of stimulated emission via DPPs.

1.4.2 Device Operation

Figure 1.7a–c shows light emitted from the fabricated Si-LED (forward-bias voltage 7 V, current density $2\,A/cm^2$) taken with a band-filtered visible CCD camera at room temperature. They clearly reveal that the light emission spectrum from the Si-LED contained all three primary colors: blue, green, and red, with photon energies of 3.1, 2.1, and 2.0 eV (wavelengths 400, 590, and 620 nm), respectively. This confirms that the fabricated Si-LED showed light emission in the visible region when a forward-bias voltage was applied.

Curves A and B in Fig. 1.8 are light emission spectra of the devices at room temperature before and after the DPP-assisted annealing, respectively. No light emission was observed before annealing. However, after the annealing, a broad emission spectrum was observed. Noticeably, there are three dominant peaks in this spectrum: the first one at 3.1 eV (400 nm wavelength), which corresponds to blue, and the other two close peaks at photon energies of 2.1 eV (590 nm wavelength) and 2.0 eV (620 nm wavelength), which correspond to green and red, respectively.

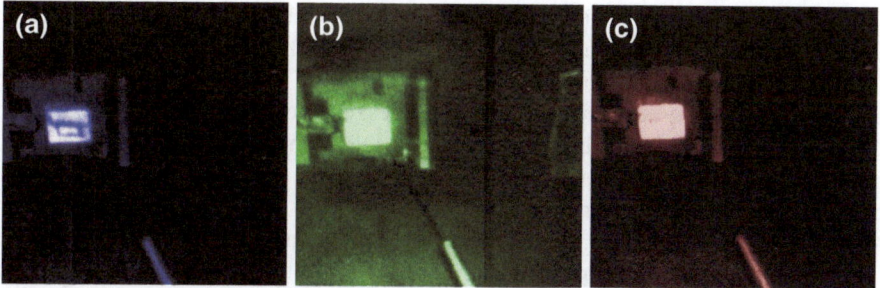

Fig. 1.7 Light emitted from the Si-LED. **a–c** are *blue*, *green*, and *red light* with photon energies of 3.1, 2.1, and 2.0 eV (wavelength; 400, 590, and 620 nm), respectively

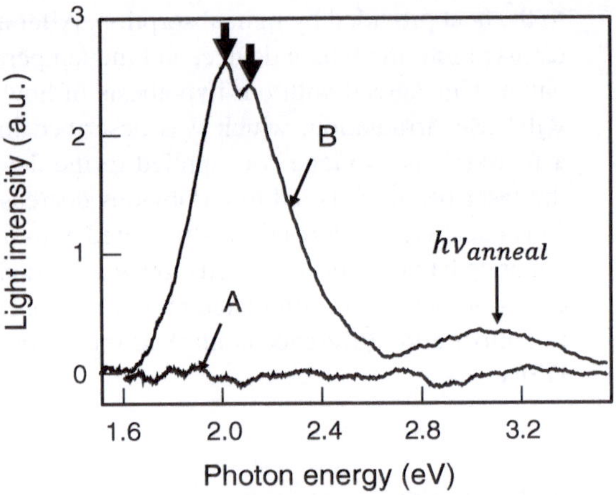

Fig. 1.8 The light emission spectra of the devices at room temperature. *Curves A and B* are the spectra taken before and after the DPP-assisted annealing, respectively. *Two downward arrows* represent the spectral peaks at 2.0 (620 nm-wavelength) and 2.1 eV (590 nm-wavelength)

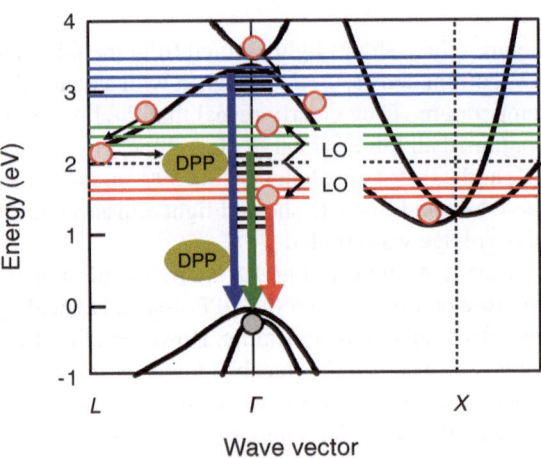

Fig. 1.9 The band structure of Si and the phonon-coupled electronic levels. The *blue, green*, and *red horizontal lines* represent phonon levels involved in the light emission in the visible region. Three *blue, green*, and *red downward arrows* represent the radiative relaxation processes of electrons from the L-point to the ground state

The above light emission characteristics agree with the principle described in Sect. 3.1 and will be explained in detail here. Figure 1.9 shows the band structure of Si (black curves) and the phonon-coupled electronic levels $|E_{ex}; el\rangle \otimes |E_{ex}; phonon\rangle$ (in particular, the blue, green, and red horizontal lines represent phonon levels involved in the light emission in the visible region). Since a DP strongly couples with phonons, a transition between the phonon-coupled electronic levels takes place if the generation probability of a DP that is resonant with the transition energy is sufficiently high. After the annealing process, almost all of the B domains in the Si become suitable for generating DPPs whose photon energy corresponds to the light irradiated during annealing. Because the photon energy $h\nu_{anneal}$ of the annealing light was 3.1 eV (400 nm wavelength), high-energy excited electrons at the Γ-point of the conduction band relaxed to the ground state via 3.1 eV-photon emission (blue downward arrow in Fig. 1.9), resulting in the spectral peak at 3.1 eV in the emission spectrum. This photon emission is the result of photon breeding, as was described above.

Moreover, the injected electrons also tended to relax to the lower energy level via intra-band relaxation, i.e., to the L- and X-points in the conduction band. Therefore, the density of electrons at those points in the conduction band was considerably high. Because the emission intensity is also proportional to the number of electrons, the radiative transition of electrons from the L-point resulted in light emission in the proximity of 2.0 eV.

The broad spectrum of the observed light emission is a result of the interaction of electrons with phonons. Since the probability of this interaction is inversely proportional to the number of phonons involved in phonon absorption or emission, the emission spectrum has two peaks at photon energies of 2.0 and 2.1 eV (two downward arrows on the curve B in Fig. 1.8). These photon energies are close to the energy level of the L-point, indicating the interaction with one longitudinal optical (LO)-mode phonon in the light emission process. Since there is a large difference in momentum between an electron at the L-point of the conduction band and a hole at the Γ-point of the valence band, the electron at the L-point needs to interact with a phonon in order to lose or gain the momentum for emitting a photon. As a result, photons are emitted from two separate energy levels (green and red downward arrows in Fig. 1.9). This explains the existence of two peaks in the vicinity of the L-point, at 2.1 and 2.0 eV. The energy difference between the two peaks is 100 meV, which is approximately equivalent to twice the LO phonon energy [18]. The dip between the two peaks corresponds to the zero-phonon line. On the other hand, no similar peaks appeared in the spectrum at 3.1 eV. This is because the electrons directly relaxed to the valence band (the blue downward arrow in Fig. 1.9), and thus, the interactions with phonons were weaker.

As discussed above, the visible light emission is a result of the high-energy excited electrons, which are provided by the electrical power source. Therefore, it is necessary to study the dependence of the light emission performance of the Si-LED on the forward-bias voltage. Curves A and B in Fig. 1.10 show the dependences of the

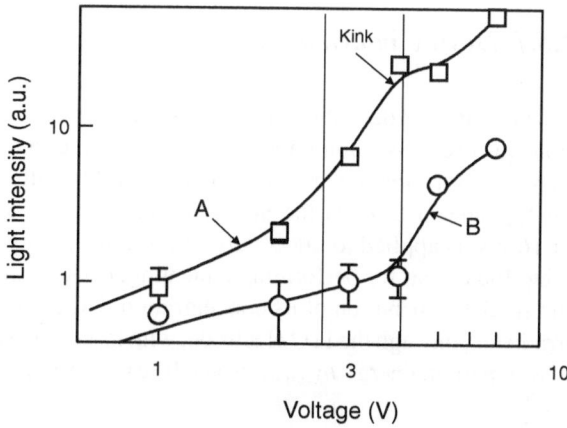

Fig. 1.10 The dependences of the height of the peaks in Fig. 1.8 on the applied voltage. *Curves A* and *B* are at 2.0 (*red light*) and at 3.1 eV (*blue light*) in Fig. 1.8, respectively

height of the peaks at 2.0 (red light) and at 3.1 eV (blue light) in Fig. 1.8 on the applied voltage. The peak at 2.1 eV (green light) showed behavior similar to the peak at 2.0 eV and is therefore omitted here. Figure 1.10 shows that, at a voltage of 2.5 V, curve A starts rising with a higher slope. This corresponds to the threshold for red light emission. Furthermore, at a voltage of about 4.0 V, curve B changes its slope. This kink corresponds to the start of blue light emission. Interestingly, a kink in curve A is also observed in the curve B.

This characteristic dependence of the emission intensity on the voltage is well explained by the emission mechanism discussed above: Since the voltage loss due to the Schottky barrier and parasitic circuit resistance is about 1.0 V, the highest energy of injected electrons is about 2.0 eV when a voltage of 3.0 V is applied. Thus, the number density of electrons that have relaxed to the L-point in the conduction band (energy level of about 2.0 eV) starts increasing. Therefore, the emission with photon energies of 2.0 and 2.1 eV appears. This transition corresponds to the change in slope of curve A in Fig. 1.10. When a sufficiently high forward bias is applied, the energy of injected electrons becomes larger than 3.1 eV, and the number density of electrons at the Γ-point increases, resulting in the appearance of emission from a transition with a photon energy corresponding to blue light. This threshold voltage corresponds to the kink in the slope of curve B. Here, the appearance of a kink in curve A is evidence for the fact that some of the injected electrons recombine to cause blue light emission, and so the number density of electrons that have relaxed to the L-point becomes relatively lower and results in the decreased emission intensity of the red light.

1.5 Infrared Light Emitting Diodes

This section reviews the fabrication and operation of a Si-LED that emits infrared light [19]. In contrast to the single-step spontaneous and stimulated emission processes described in section , those of infrared light emission are two-step processes (Sect. 1.3.2). This is because the emitted photon energy is lower than E_g.

1.5.1 Device Fabrication

As with the case described in Sect. 1.4.1, an n-type Si crystal with low As concentration was used. By doping the crystal with B, the Si crystal surface was transformed to p-type, forming a p–n homojunction. An ITO film and an Al film were deposited on opposite surfaces of the Si crystal for use as electrodes. A forward bias voltage of 16 V was applied to inject current (current density of 4.2 A/cm^2) in order to generate Joule-heat for performing annealing, causing the B to be diffused and varying the spatial distribution of its concentration. During the annealing, the Si crystal was irradiated, through the ITO electrode, with laser light (light power density, 10 W/cm^2) whose photon energy hv_{anneal} was 0.95 eV (1.30 μm wavelength).

Since $h\nu_{anneal}$ is lower than E_g of Si, the radiated light is not absorbed by the Si crystal. Therefore, in the regions where DPPs are hardly generated, B diffuses simply due to the Joule heat of the applied electrical energy. However, in the regions where DPPs are easily generated, the thermal diffusion rate of the B becomes smaller via the following processes:

(1) Since the energy of the electrons driven by the forward-bias voltage (16 V) is higher than E_g, the difference $E_{Fc} - E_{Fv}$ between the quasi Fermi energies in the conduction band E_{Fc} and the valence band E_{Fv} is larger than E_g. Therefore, the Benard–Duraffourg inversion condition is satisfied. Furthermore, since $h\nu_{anneal} < E_g$, this light propagates through the Si crystal without absorption and reaches the p–n homojunction. As a result, it generates DPPs efficiently around the domain boundaries of the inhomogeneous distribution of B. Since stimulated emission takes place via DPPs (Sect. 3.2), the electrons generate photons by the stimulated emission and are de-excited from the conduction band to the valence band via the phonon energy level.

(2) The annealing rate decreases because a part of the electrical energy for generating the Joule-heat is spent for the stimulated emission of photons. As a result, at the regions where the DPPs are easily generated, the shape and dimensions of the B inhomogeneous domain boundaries become more difficult to change.

(3) Spontaneous emission occurs more efficiently at the regions in which the DPPs are easily generated because the probability of spontaneous emission is proportional to that of stimulated emission. Furthermore, with temporal evolution of process (2), the light from stimulated and spontaneous emission spreads through the whole Si crystal, and as a result, process (2) takes place in a self-organized manner throughout the entire volume of the Si crystal.

It is expected that this method of annealing will form the optimum spatial distribution of the B concentration for efficient generation of DPPs, resulting in efficient device operation. Figure 1.11 shows the temporal evolution of the temperature of the device surface as annealing progressed. After the temperature rapidly rose to 154 °C, it fell and asymptotically approached a constant value (140 °C) after

Fig. 1.11 The temporal evolution of the temperature of the device surface as annealing progresses

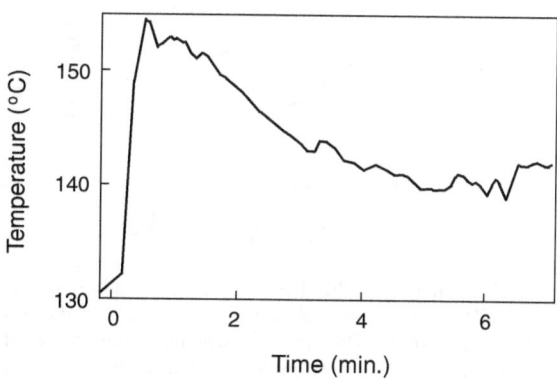

6 min, at which time the temperature inside the device was estimated to be about
300 °C. The features of this temporal evolution are consistent with those of the
principle of annealing under light irradiation described above: The temperature
rises due to the Joule-heat by the applied electrical energy. However, the temperature gradually falls because stimulated emission is induced by the DPPs
generated at the domain boundary of the inhomogeneous distribution of the B
concentration. Finally, the system reaches the stationary state.

1.5.2 Device Operation

Figure 1.12a shows the relationship between the forward-bias voltage (V) applied to
the device and the injection current (I) for a Si crystal with a surface area as large
as $9\,mm^2$ and a thickness of $650\,\mu m$. This figure indicates negative resistance at
$I > 50\,mA$, and the breakover voltage V_b was 63 V [20]. This was due to the spatially
inhomogeneous current density and the generation of filament currents, as shown in
the inset of Fig. 1.12a. In other words, the B distribution had a domain boundary,
and the current was concentrated in this boundary region. A center of localization
where the electrical charge is easily bound was formed in this current concentration
region, and a DPP was easily generated there. That is, the negative resistance is consistent with the principle of the device fabrication described in Sect. 1.5.1. The reason
why V_b was higher than the built-in potential of the Si p–n junction is because of the
high total resistance due to the thick Si crystal wafer and the large contact resistance
between the electrodes and the Si crystal wafer. In addition, although the device
surface temperature during annealing (Fig. 1.11) was too low for diffusing the B,
localized heating occurred due to the filament currents described above, which made

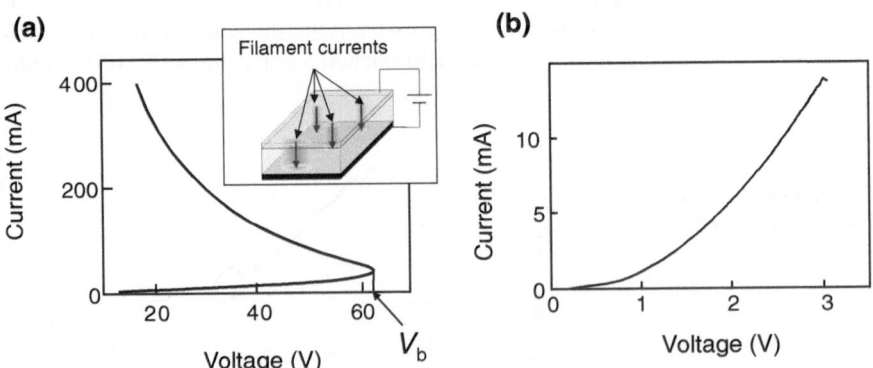

Fig. 1.12 The relationship between the forward-bias voltage (V) to the device and the injection
current (I) for the Si crystal. **a** The surface area of the Si crystal is as large as $9\,mm^2$ and the
thickness is $650\,\mu m$. The *inset* is a schematic explanation of the filament currents generated due
to the spatially inhomogeneous current density. **b** The surface area is as small as $0.6\,mm^2$, and the
thickness is $120\,\mu m$

Fig. 1.13 Photographs of the device. **a** and **b** were taken without and with current injection, respectively, at room temperature

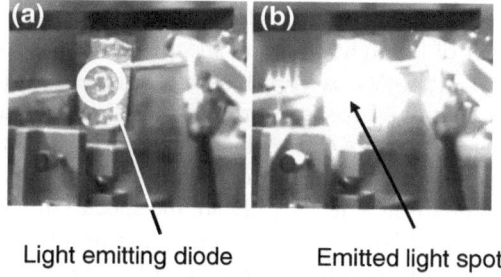

Light emitting diode   Emitted light spot

the temperature inside the device sufficiently high. From secondary ion-microprobe mass spectrometry, it was confirmed that the B penetration depth was increased to at least 300 nm by the annealing. For comparison, when the device had a size as small as 0.6 mm^2 in area and 120 µm in thickness, the $V - I$ characteristic did not exhibit such negative resistance but was a $V - I$ characteristic identical to that of an ordinary LED using a direct transition-type semiconductor, as shown in Fig. 1.12b.

Figure 1.13a, b are photographs showing the device without and with current injection (current density, 4.2 A/cm^2), respectively, at room temperature, which were taken by an infrared CCD camera (photosensitive bandwidth: 1.73–1.38 eV, wavelength: 0.90–1.70 µm) under fluorescent lamp illumination. Figure 1.13b shows a bright spot of light with an optical power as high as 1.1 W, which was emitted by applying 11 W of electrical power.

It should be pointed out that a conventional Si photodiode can emit light even though its efficiency is extremely low. Figure 1.14a shows the light emission spectrum of a commercial photodiode (Hamamatsu Photonics, L10823) at an injection current

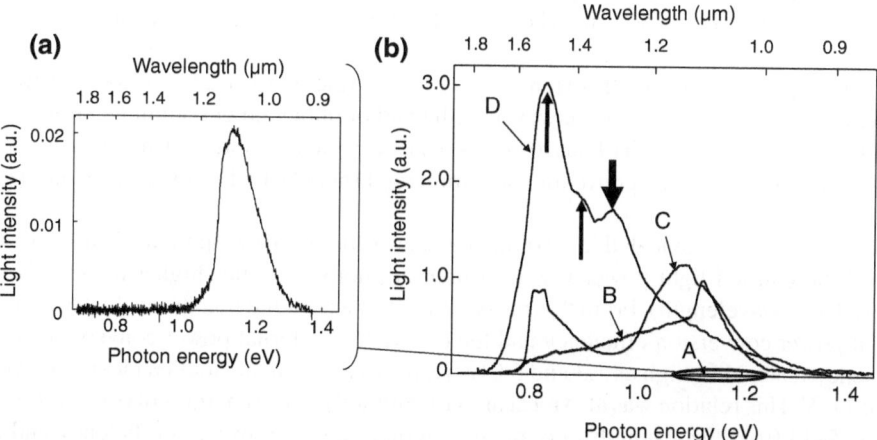

Fig. 1.14 The light emission spectra. **a** The spectrum of a commercial photodiode (Hamamatsu Photonics, L10823) at an injection current density of 0.2 A/cm^2. **b** *Curve A* is identical to the curve in (**a**). *Curves B–D* are the spectra of the devices fabricated by annealing for 1, 7, and 30 min, respectively

density of $0.2\,A/cm^2$. Higher current injection damaged the photodiode. The main part of the light emission spectrum in this figure is located at energies higher than $E_g(=1.12\,eV)$, as a result of the indirect transition caused by phonon scattering [21].

Figure 1.14b shows the light emission spectra of three devices at an injected current density of $1.5\,A/cm^2$, which were fabricated by the method described in the previous section. Curve A in this figure is identical to the curve in Fig. 1.14a, drawn for comparison. Curves B–D are the spectra of the devices fabricated by annealing for 1, 7, and 30 min, respectively. Their profiles considerably differ from that of curve A, and the spectra are located at lower energies than E_g. The light intensity values of the curves B–D appear low at energies lower than 0.8 eV, which is due to the low sensitivity of the photodetector used for the measurement. Although the light emission spectrum of the device fabricated using 1 min of annealing (curve B) still had a clear peak around E_g, the spectrum broadened and reached an energy of 0.75 eV (1.65 μm wavelength). That of the device annealed for 7 min (curve C) showed a new peak at around 0.83 eV (1.49 μm wavelength). In the case of the device annealed for 30 min (curve D), no peaks were seen around E_g; instead, a new peak appeared, identified by a downward thick arrow, at an energy that corresponds to the photon energy $h\nu_{anneal}$ (=0.95 eV, 1.30 μm wavelength) of the light radiated in the annealing process. This peak is evidence that DPPs were generated by the light irradiation and that the B diffusion was controlled. In other words, photon breeding took place also by the two-step process (Sect. 1.3.2). The value of the emission intensity at the highest peak (identified by the left thin upward arrow) on curve D was 14-times and 3.4-times higher than those of the peaks on curves B and C, respectively.

Here, the separations between the energies identified by two upward thin arrows (0.83 and 0.89 eV), and by the downward thin arrow (0.95 eV) were 0.06 eV, which is equal to the energy of an optical phonon in Si. This means that the two upward thin arrows show that the DPP with an energy of 0.95 eV was converted to a free photon after emitting one and two optical phonons. This conversion process demonstrates that the light emission described here uses the phonon energy levels as an intermediate state.

The spectrum of curve D extended over the energies 0.73–1.24 eV (1.00–1.70 μm wavelength), which covers the wavelength band of optical fiber communication systems. The spectral width of curve D was 0.51 eV, which is more than 4-times greater than that (0.12 eV) of a conventional commercial InGaAs LED with a wavelength of 1.6 μm.

For the device annealed for 30 min, the relation between the applied electric power and the emitted light power was measured at photon energies higher than 0.73 eV (1.70 μm wavelength). From this measurement, it was found that the evaluated external power conversion efficiency and the differential external power conversion efficiency reached as high as 1.3 and 5.0 %, respectively, with an applied electric power of 11 W. This relation was also measured for emitted photon energies of 0.11–4.96 eV (0.25–11.0 μm wavelength), giving an external power conversion efficiency and a differential external power conversion efficiency as high as 10 and 25 %, respectively.

In order to estimate the quantum efficiency, the relation between the injected current density I_d and the emitted light power density P_d at photon energies higher than

Fig. 1.15 Progress in external quantum efficiency. The *solid curve* is the logistic curve fitted to the experimental values. The *gray thick horizontal line* represents the value for a commercially available InGaAs near-infrared LED

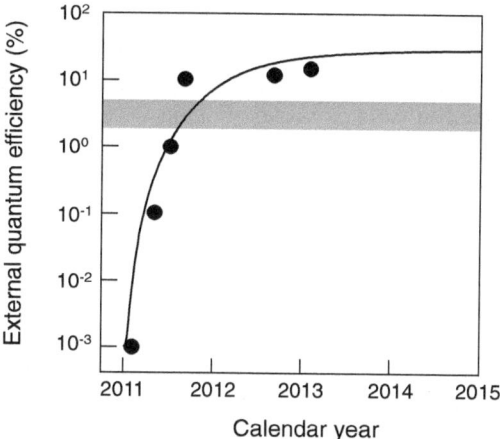

0.73 eV (1.70 μm wavelength) was measured for the device annealed for 30 min. From this measurement, it was found that P_d was proportional to I_d^2, which is due to the two-step spontaneous emission process; i.e., one electron is converted to two photons. For comparison, P_d is proportional to I_d in a conventional LED device. Furthermore, the external quantum efficiency was estimated to be 15 % at $I_d = 4.0$ A/cm^2, and the differential external quantum efficiency was estimated to be 40 % at $I_d = 3.0$–4.0 A/cm^2. From the measured relation between I_d and P_d at photon energies of 0.11–4.96 eV (0.25–11.0 μm wavelength), the external quantum efficiency at photon energies higher than 0.73 eV (1.70 μm wavelength) was as high as 150 %. The reason why this value is higher than 100 % is that the two-step spontaneous emission process converts one electron not to one photon but to two photons.

Figure 1.15 shows the progress made in increasing the external quantum efficiency in the wavelength range of 1.32 ± 0.15 μm. By early 2013, an efficiency of 15 % had been reached. This value is already higher than those of commercially available 1.3 μm-wavelength LEDs using the direct-transition-type semiconductor InGaAs, such as Hamamatsu Photonics L7866 and L10822 (external quantum efficiencies of 2 and 5 %, respectively). The solid curve in this figure is the least-squares fitted logistic curve, which has been popularly used to represent population growth, technological progress, and increases in the number of articles on the market. This curve fits the experimental values well, showing the rapid, smooth progress of Si-LED technology.

1.6 Strength of Phonon Coupling

The main features of the Si-LEDs reviewed in Sects. 1.4 and 1.5 are that the emitted photon energy does not depend on E_g; instead, the emitted light has approximately the same photon energy as the photon energy $h\nu_{anneal}$ of the light radiated during annealing. In other words, photon breeding takes place. Also, phonon sidebands are found in the light emission spectrum, and these are caused by the creation of DPPs. The phonons constituting these DPPs have been theoretically shown to be multi-mode coherent phonons [22]. The parameter representing the coupling strength between electron–hole pairs and phonons is called the Huang–Rhys factor, S [23]. The value of S can be obtained experimentally from the intensities of the phonon sidebands [24, 25]. When a DPP is created, an increase in the value of S is expected, but its magnitude remains unknown.

This section estimates the value of S in the fabricated Si-LEDs by comparing the measured emission intensities of the phonon sidebands found in its emission spectrum and a simulation result of the light emission process [26]. In the case of the device used in this section, the photon energy of the light absorbed in the depletion layer is 1.2–3.0 eV, which was estimated based on the depletion layer depth and thickness obtained from the B doping conditions. There is no singular point in the Si band structure in this range of energies and especially in the range corresponding to the photon energy of the emitted light (1.4 eV). Thus, this range is suitable for evaluating the involvement of phonons in the light emission process. Therefore, the surface of the device was irradiated with light having a photon energy $h\nu_{anneal}$ of 1.4 eV during annealing. The principle of light emission and DPP-assisted annealing are the same as those described in Sects. 1.3.1 and 1.4.1, respectively.

Ion implantation was used to dope B into an n-type Si crystal in order to form a p-type layer, thus fabricating a p–n homojunction. The highest B ion implantation energy was 700 keV, the concentration was $1 \times 10^{19}\,cm^{-3}$, and the thickness of the p-type layer was 2 μm. After dicing this crystal, an ITO film and a Cr/Al film were respectively deposited on the surfaces of the p-type and n-type layers for use as electrodes. A forward-bias voltage of 2 V was applied to the device (current density 0.96 A/cm^2) while simultaneously irradiating it with laser light ($h\nu_{anneal} = 1.4$ eV) to perform DPP-assisted annealing for 3 h.

Figure 1.16 shows the light emission spectra measured when a forward current (current density 3.2 A/cm^2) was injected into the Si-LED. The curves A and B are the spectra obtained 1 and 3 h after starting annealing, respectively. By comparing these curves, it can be seen that the emission intensity showed almost no change at photon energies of 1.25 eV and above, whereas at photon energies below 1.25 eV, the emission intensity was considerably increased. A possible reason for this is that, in the regions where low-energy photons are easily generated, the energy dissipation level due to stimulated emission produced in those regions is small, like those discussed in (2) in Sect. 1.4.1. Thus, the thermal diffusion rate is similar to that in region (1) in Sect. 4.1, and therefore, annealing proceeds slowly. This phenomenon suggests the validity of the annealing principle discussed in Sect. 1.4.1. In Fig. 1.16, the photon energy $h\nu_{anneal}$ (=1.4 eV) of the light irradiated during annealing is indicated by the upward arrow. From the curves A and B, it is found that this Si-LED emitted light with a photon energy approximately equal to $h\nu_{anneal}$; i.e., photon breeding took place. Furthermore, multiple sidebands are observed in the curve B (indicated by the downward arrows). The spacings between these downward and upward arrows are approximately constant at 60 meV, which agrees with the energy of optical phonons in Si [27], and therefore, these sidebands likely originate from phonons.

Fig. 1.16 The light emission spectra of the Si-LED. The *curves A and B* show the spectra obtained 1 and 3 h after starting annealing, respectively

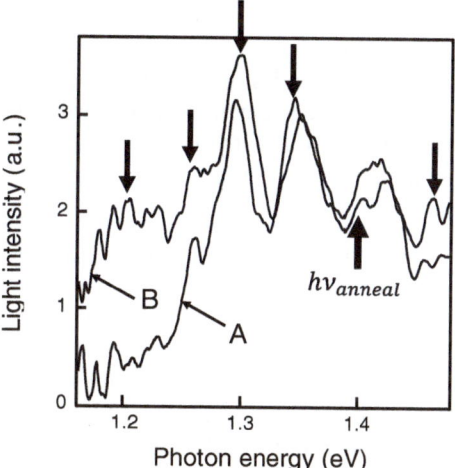

In the Si-LED fabricated here, since it involves a simple single-step transition, only one photon is emitted due to the radiative relaxation of a single electron. In addition, the emission lifetime of a Si-LED using DPPs is on the order of 1 ns (Sect. 1.10.1), which is considerably shorter than the lifetime of the weak light emission from a normal Si bulk crystal. However, it is three to four orders of magnitude larger than the intraband relaxation time of electrons in Si (0.1–1 ps) [28, 29]. Therefore, it is reasonable to assume that the injected electrons do not radiatively relax during the relaxation process α in Fig. 1.17, but instead radiatively relax by coupling with phonons after they have relaxed to the bottom of the conduction band, as shown by process β. This indicates that the electron–hole pairs have an energy identical to E_g. Therefore, it can be concluded that the difference between the photon energy

Fig. 1.17 Schematic explanation of the light emission process via DPP energy levels. α and β represent two possible relaxation processes

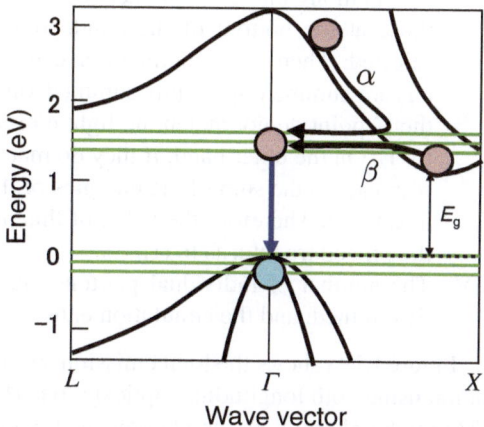

of the emitted light and E_g is the energy of the phonons that have coupled with the electron–hole pairs.

The light emission spectrum was obtained via simulation, and the coupling strength between electron–hole pairs and phonons was examined by comparing the simulation results with the measured spectrum. The following four assumptions were made in the simulation:

(1) The electron and hole are at the bottom of the conduction band and at the top of the valence band, respectively, and the momentum and energy of the phonon are imparted to this electron–hole pair. This assumption was made based on the fact that the electron and hole intraband relaxation time is shorter than the emission lifetime via the phonon level, as described above.
(2) The electron–hole pair couples with coherent phonons including all optical modes. This was assumed based on experimental results [30] and theoretical considerations [22] regarding DPPs.
(3) The probability distribution of the number of phonons that couple with an electron–hole pair follows a Poisson distribution. In other words, the probability of n phonons coupling with an electron–hole pair is proportional to $S^n/n!$ This means that coupling of electron–hole pairs and phonons via DPPs occurs randomly.
(4) Radiative relaxation occurs by following the momentum conservation law between the initial state and the final state.

The simulation procedure is as follows:

(I) Based on assumption (3), the number of phonons that couple with an electron–hole pair is determined.
(II) Based on assumption (2), from among all of the optical phonons, phonons whose number was determined in (I) are randomly selected. From this number and the dispersion relation of phonons in Si [31], the momentum and energy of the coupling phonons are determined.
(III) From among the electron energy levels between the Γ point and the X point, those at the bottom of the conduction band, having the lowest energy are selected. Then, their momenta and the momenta of the phonons selected in (II) are summed up. If this summed value and the value of the momentum at the Γ point do not match, no light emission occurs, and the procedure returns to (I). On the other hand, if they do match, the light whose photon energy corresponds to the sum of the energies of the electron–hole pair and the phonons is emitted. Therefore the value of this photon energy can be calculated. Then the procedure returns to (I).
(IV) The number of individual photons, obtained by repeating (I)–(III) above, is determined, and the simulation ends.

Figure 1.18a shows the light emission spectrum obtained by performing the simulation using both longitudinal optical mode (LO mode) and transverse optical mode (TO mode) phonons as the phonons that couple with the electron–hole pair, where n is the number of phonons. Multiple peaks corresponding to each n are observed

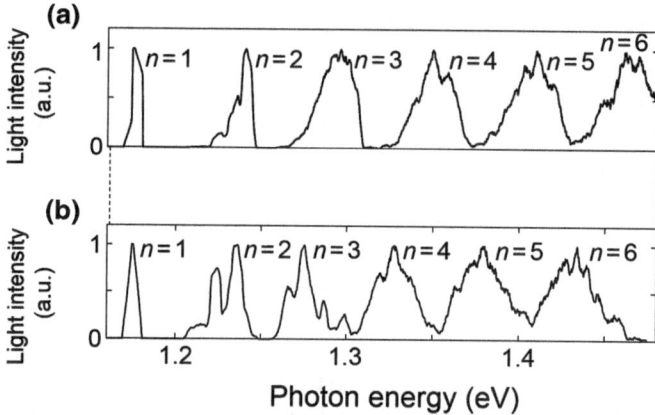

Fig. 1.18 The light emission spectra obtained by performing the simulation. The parameter n is the number of phonons that coupled with an electron–hole pair via a DPP. **a** The results obtained by using both longitudinal optical mode (LO-mode) and transverse optical mode (TO-mode) phonons as the phonons that couple with the electron–hole pair. **b** The result obtained by using only LO-mode phonons

in this spectrum, similarly to the measured results (Fig. 1.16). For ease of viewing the spectral shape, the heights of all peaks are normalized to unity. By comparing the curve in this figure with the curve shown in Fig. 1.16, it can be concluded that the phonons that couple with the electron–hole pair are composed of both LO-mode and TO-mode phonons since the photon energies at the peaks match, and since the spectral shapes resemble each other. For the sake of comparison, Fig. 1.18b shows a light emission spectrum obtained by performing the simulation using only LO-mode phonons. The positions of the peaks on the curve in Fig. 1.18b differ from the positions of the peaks in Figs. 1.16 and 1.18a, from which it is confirmed that both LO-mode and TO-mode phonons are involved in the coupling with the electron–hole pair.

Next, to determine the value of the Huang–Rhys factor, S, the heights of the emission peaks were considered. As described in assumption (3) above, since the distribution of the number of phonons n follows a Poisson distribution that depends on the value of S, the height of each spectral peak in Fig. 1.18a also depends on the value of S. Hence, least squares fitting was carried out on the heights of each peak in Fig. 1.18a and each peak on the curve B in Fig. 1.16. As a result, the value of S was found to be 4.08 ± 0.02. In a normal bulk Si crystal, on the other hand, S ranges from 0.001 to 0.01 [32], which means that the DPP-mediated coupling between the electron–hole pair and phonons is two to three orders of magnitude stronger. The peak widths in the light emission spectrum of the actual device were found to differ from those in the simulated spectrum. This is because coupling between the electron–hole pair and acoustic phonons was not taken into account in the simulation.

1.7 Contribution of the Multimode Coherent Phonons

Previous sections have revealed that DPs are accompanied by multimode coherent phonons, resulting in the creation of DPPs. High-resolution Raman spectroscopy [33] and Fourier transform infrared spectroscopy [34] have been used to study the spectral properties of phonons generated in a super-wavelength sized space in a macroscopic material. Furthermore, coherent phonons (CPs) have been generated by irradiating macroscopic materials with short optical pulses, and their spectral properties have been studied by measuring CP-induced temporal variations in macroscopic physical quantities of the materials, such as the optical reflectivity [35, 36]. Theoretical studies have also been carried out to explain several phenomena caused by these CPs [37]. In contrast to these conventionally studied CPs, the multimode coherent phonons are excited by the DPs in nanometric spaces. These multimode coherent phonons in the nanometric spaces (MCP-NSs) have large uncertainties in their momenta (wavenumbers) because of the sub-wavelength size of the nanometric spaces in which they are generated. In this section, the origin of the sidebands observed in the light emission spectrum of a visible Si-LED is reviewed based on the results of measurement of optical phonons in the MCP-NSs by using pump–probe spectroscopy [38].

A Si-LED was fabricated by applying a forward-bias voltage of 3.5 V (injected current: 240 mA) while irradiating the Si crystal with pulsed laser light for 1 h. The laser light had a photon energy $h\nu_{anneal}$ of 3.1 eV (400 nm wavelength), a pulse width of 100 fs, a repetition frequency of 80 MHz, a pulse energy density of 5×10^{-8} J/cm^2, and an average power density of 4 W/cm^2. The principles of light emission and DPP-assisted annealing are the same as those of Sects. 3.1 and 4.1, respectively. The reason for using a pulsed laser, not a CW laser, is to maintain a sufficiently high peak power for efficient annealing by light absorption.

Figure 1.19 shows the main part of the spectrum of the light emitted from the fabricated Si-LED with an injected current of 450 mA. The curve in this figure clearly shows three peaks in the visible region. The spectral component at 3.1 eV ($=h\nu_{anneal}$), which originated from the light irradiated in the annealing process (i.e., photon breeding), is not displayed due to the low efficiency of the diffraction grating in the monochromator used for the spectral measurement. The separations between adjacent peaks are about 250 meV (60 THz), from which these peaks can be identified as phonon sidebands originating from the strong coupling between DPs and MCP-NSs in the Si crystal, as will be discussed in the following.

In order to generate and evaluate the MCP-NSs for studying the origin of the sidebands in Fig. 1.19, pump–probe laser spectroscopy was employed based on the principle of impulsive stimulated Raman scattering (ISRS) [39]. If the spectral width of the optical pulse (carrier frequency ν) from the light source is wider than the eigenfrequency of the phonon, ν_p, the intensity of the frequency components $\nu - \nu_p$ in the optical pulse can be maintained sufficiently high for generating the MCP-NSs efficiently.

Since the generated MCP-NSs temporally modulate the Coulomb potential of the interaction with electrons, the energy band structure of the electrons is modulated, resulting in modulation of the optical reflectivity of the Si crystal [36, 40]. Among

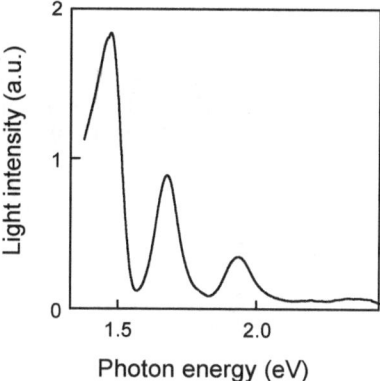

Fig. 1.19 The main part of the spectrum of the light emitted from the Si-LED with an injected current of 450 mA

the coherent and incoherent phonons, this modulation enables selective detection of the MCP-NSs because the thermally excited incoherent phonons do not induce any variations in the optical reflectivity.

The temporal variation of the optical reflectivity can be acquired by plotting the reflected probe beam intensity as a function of the time difference, t, between the arrival times of the incident probe and pump optical beams at the Si crystal surface. A Ti:sapphire laser was used as a light source. The center wavelength was 780–805 nm (photon energy, 1.54–1.59 eV), and the pulse width was 15 fs. The temporal variation of the optical reflectivity was measured by lock-in detection of the probe beam intensity reflected from the Si crystal surface.

Figure 1.20a shows a normalized fractional variation of the optical reflectivity measured as a function of the time difference t defined above. The curve in this

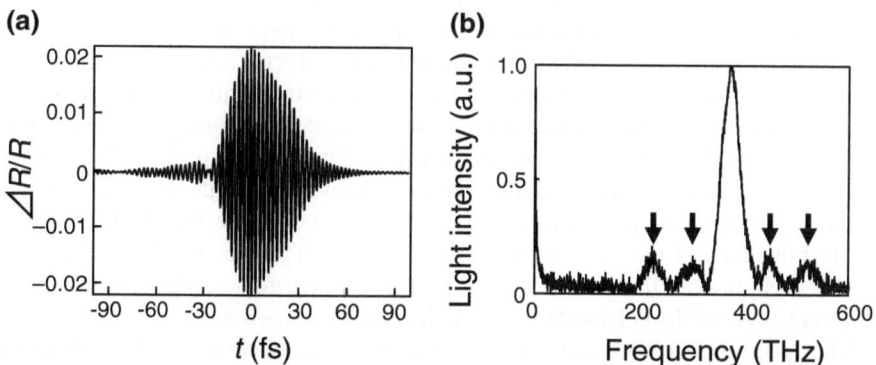

Fig. 1.20 Measured optical reflectivity. **a** Normalized fractional variation of the optical reflectivity measured as a function of the time difference t between the arrival times of the incident probe and pump optical beams at the Si crystal surface. **b** The spectrum obtained by Fourier transforming the fractional variation of the optical reflectivity

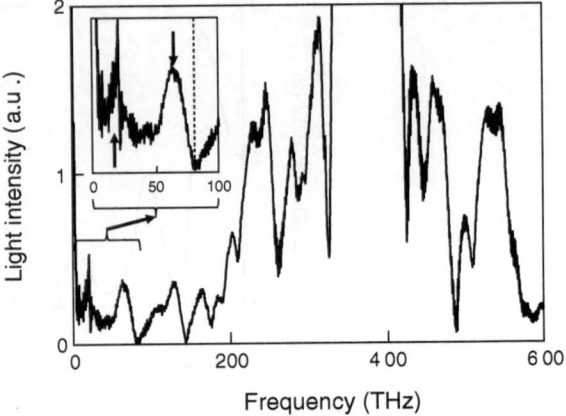

Fig. 1.21 The measured Fourier-transformed spectrum. The inset represents the magnified spectral curve of the signals originating from the MCP-NSs

figure represents a rapidly oscillating, amplitude-modulated optical interference signal between the pump and probe beams. Figure 1.20b shows the spectrum obtained by Fourier transforming the fractional variation of the optical reflectivity when the average powers of the pump and probe beams were 120 and 20 mW, respectively. Their beam spot diameters were 1 mm. As represented by the four downward arrows, the spectral curve shows several sidebands on both sides of the high spectral peak of the optical interference signal at 385 THz. These sidebands originate from the amplitude and phase modulations of the reflected probe beam intensity, i.e., the modulation of the optical reflectivity of the Si crystal. These results suggest that there is strong coupling between the MCP-NSs and the pump-beam photons.

In order to evaluate the spectral properties in Fig. 1.20b more quantitatively, the average power of the probe beam was increased to 120 mW to increase the measurement sensitivity. Figure 1.21 shows the measured Fourier-transformed spectrum. In this curve, the interference signal peak at 385 THz is much higher than the maximum on the vertical axis. On both sides of this over-scaled peak, many more sidebands are seen, as compared with Fig. 1.20b. In addition to the optical interference signal and the sidebands in Figs. 1.20b and 1.21 also reveals the unique signals originating from the MCP-NSs. The spectral curve of these signals is magnified and shown in the inset of Fig. 1.21. These signals are manifested by spectral peaks at frequencies lower than 80 THz, as represented by the two arrows. The high-frequency cut-off, which is governed by the value $\nu - \nu_p$, was confirmed to be 80 THz because the spectral intensity of the curve in the inset decreased to zero at 80 THz (represented by a vertical broken line in the inset). The curve in the region higher than 80 THz represents tails of the sidebands of the optical interference signal.

The spectral peaks shown by the two arrows on the curve in the inset are at the frequencies of 18 and 64 THz (74 and 265 meV). It is easily seen that the spectral intensity at 64 THz, which is defined by the area under the bell-shaped spectral curve, is larger than that at 18 THz. As a result of the comparison between the spectral intensities, it was found that the sidebands with a separation of about 60 THz in Fig. 1.19 originated from the spectral peak at 64 THz in this magnified curve.

For a more detailed evaluation of the curve in the inset, it was fitted by the superposition of the Lorentzian spectral curves of the fundamentals and harmonics of the LO-mode phonons with the center frequency $n\nu_p$, which can contribute to forming the MCP-NSs. The integer n represents the order of the harmonics, where $n = 1$ represents the fundamental. The center frequency ν_p was set to 15.6 THz for the LO-mode phonons by referring to previous experimental and theoretical studies. (From a study of the LO-mode CPs in a Si crystal, the eigen-frequency of the LO-mode phonon has been measured to be 15.6 THz [35, 36].) Since TO-mode phonons can also contribute to forming the MCP-NSs, these could also be measured; however, with pump–probe spectroscopy, only phonons with a wavenumber of 0 (the Γ point in wavenumber space) are measured and, since the frequencies of the LO-mode phonons and TO-mode phonons at the Γ point are approximately equal [41], it is not possible to distinguish between them. The full-width at half maximum of the Lorentzian spectral curve was fixed at 10 THz (41 meV) by referring to the average phonon energy (30 meV) and additional scattering by incoherent phonons at room temperature [35, 42].

The broken curve A in Fig. 1.22 represents the result of the fitting, whereas the solid curve B is a copy of the curve in the inset of Fig. 1.21. The broken curves C–E represent the Lorentzian spectral curves of the fundamental, and the second- and fourth-order harmonics of the LO-mode phonons, respectively, whose superposition yields the broken curve A. The contribution from the third-order harmonics was negligibly small. As a result of this accurate fitting of the broken curve A to the curve B, the spectral peak at 64 THz in the inset of Fig. 1.21 was found to be composed of the fourth-order harmonic of the LO-mode phonons. The spectral peak at 18 THz is composed of the fundamental of the LO-mode phonons. The second- and third-order harmonics of the LO-mode phonons contribute to the lower signal. Since the spectral intensity of the 64 THz peak was the largest, it was found that the fourth-order harmonic of the LO-mode phonons couple most strongly with the pump-beam photons and are the origin of the sidebands in the emission spectrum in Fig. 1.19.

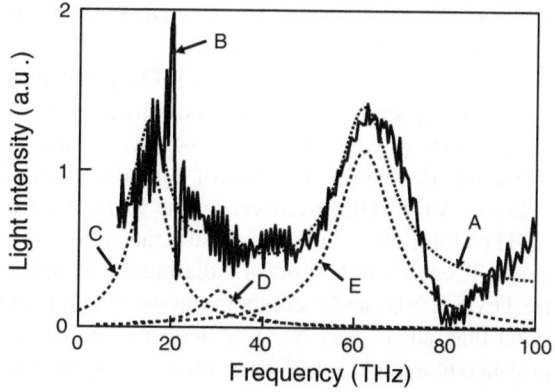

Fig. 1.22 The result of the curve fitting. *Curve A* is the result of fitting. *Curve B* is a copy of the curve in the *inset* of Fig. 1.21. *Curves C–E* represent the Lorentzian spectral curves of the fundamental and the second- and fourth-order harmonics of the LO-mode phonons, respectively

From the discussion above, the fundamental and harmonics of the LO-mode phonons were revealed to be the constituent components of the multimode coherent phonons generated in the nanometric spaces around the domain boundaries of B in the Si-LED.

1.8 Infrared Laser

Si has also been the subject of extensive research for use in fabricating lasers since it shows excellent compatibility with electronic devices [43]. For example, there are reports in the literature on Raman lasers [44] and lasers utilizing quantum size effects in Si [45]. However, parameters such as the operating temperature, efficiency, wavelength and so forth are still not adequate for practical adoption of lasers. To solve these problems, this section reviews an infrared Si lasers fabricated by using DPPs and showing continuous-wave operation at room temperature [46].

In the case where the Bernard–Duraffourg inversion condition is satisfied, optical amplification gain occurs. Furthermore, if the fabricated devices have an optical cavity structure for confining the emission energy, and if the optical amplification gain is larger than the cavity loss, there is a possibility of laser oscillation occurring as a result of stimulated emission.

An As-doped n-type Si crystal was used as a device substrate, whose electrical resistivity and thickness were $10\,\Omega\,cm$ and $625\,\mu m$, respectively. This crystal was doped with B by ion implantation to form a p-type layer. The implantation energy for the B doping was $700\,keV$, and the ion dose density was $5 \times 10^{13}\,cm^{-2}$. After forming a p–n homojunction, an ITO film was deposited on the p-layer side of the Si substrate, and an Al film was deposited on the n-substrate side for use as electrodes. Next, the Si crystal was diced to form the device. The device area was about $400\,mm^2$. Similarly to the device described in Sect. 5.1, the substrate was irradiated with infrared laser light having a photon energy $h\nu_{anneal}$ of $0.94\,eV$ ($1.32\,\mu m$ wavelength) and a power density of $200\,mW/cm^2$, during which annealing was performed by applying a forward-bias current of $1.2\,A$ to generate Joule-heat, causing the B to diffuse. With this method, the spatial distribution of the B concentration changes, forming domain boundaries in a self-organized manner, which have shapes and distributions suitable for efficiently inducing the DPP-assisted process.

Next, to fabricate a Si laser, the ITO electrode and the Al electrode were removed by etching. Then, a ridge waveguide structure was fabricated by using conventional photolithography. A SiO_2 film, used as a mask in wet chemical etching of Si by KOH, was deposited by means of tetraethyl orthosilicate chemical vapor deposition (TEOS-CVD). After transferring the mask pattern ($10\,\mu m$ linewidth) to the SiO_2, KOH etching was conducted to fabricate a ridge waveguide structure with a depth of $2\,\mu m$. Then, an SiO_2 film for isolating the Si wafer and the electrode was deposited by TEOS-CVD, and a contact window was formed on top of the ridge waveguide. After that, an Al electrode was deposited by DC sputtering. The substrate was then polished to a thickness of $100\,\mu m$, and Al was deposited on the reverse side of the Si

Fig. 1.23 The device structure of a Si laser. **a** Schematic explanation of the structure. **b** Optical micrographs of a fabricated Si laser

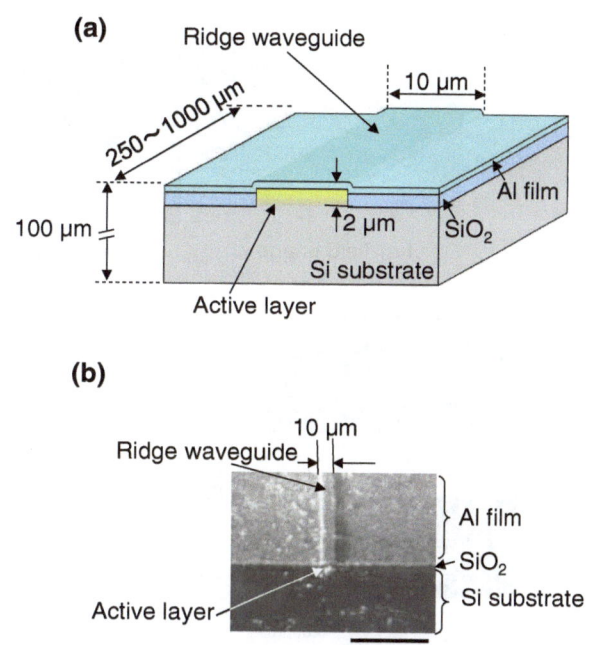

substrate. The sample was cleaved to various cavity lengths, and the cleaved facets served as cavity mirrors. Several Si laser devices were fabricated with the above method. All the experiments described below were conducted at room temperature (15–25 °C).

Figure 1.23a illustrates the device structure. Secondary ion mass spectrometry measurements confirmed that the active layer formed in the p–n homojunction was located at a depth of 1.5–2.5 μm from the surface of the Si substrate. This corresponds to the bottom of the ridge waveguide. Figure 1.23b shows optical micrographs of a fabricated Si laser. The width and thickness of the ridge waveguide constituting the cavity were 10 and 2 μm, respectively. Several Si lasers were fabricated, whose cavity lengths were 250–1,000 μm. The sum of the guiding loss in the ridge waveguide and the optical scattering loss at the cleaved facets was estimated to be 70 % for the TE-polarization component. Because this ridge waveguide does not exhibit optical confinement in the thickness direction, the guiding loss is large; it can be estimated to be 90 % or more. However, because an active layer that efficiently generates DPPs is formed in the p–n homojunction by the DPP-assisted annealing, the effective refractive index of the active layer is higher than the refractive index of the surrounding area. Therefore, the actual guiding loss was smaller than the above value. The current density dependency of the optical amplification gain, which resulted from irradiating this Si laser with 1.32 μm-wavelength laser light, was measured. From this measurement, the transparent current density was estimated to be $J_{tr} = 26.3\,\text{A/cm}^2$.

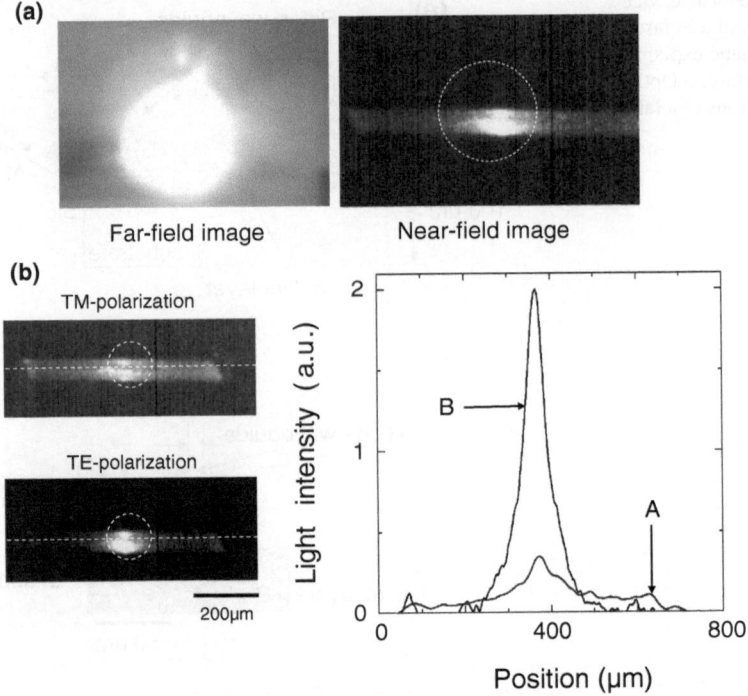

Fig. 1.24 The optical radiation patterns from the Si laser. **a** Far-field and near-field images. **b** The TM- and TE-polarization components of the near-field image in (**a**). The cross-sectional light intensity profiles are also taken along the *white dotted lines* in these images. The *curves A* and *B* are the profiles for the TM- and TE-polarization components, respectively

This value is about 1/10 that for a conventional laser device made using a direct-transition-type semiconductor [47], demonstrating that this ridge waveguide with the p–n homojunction structure has adequate performance for use as a laser.

For the laser with a cavity length of 750 μm, the far-field images of the optical radiation pattern are shown in Fig. 1.24a. A near-field image is also shown. There is a ridge waveguide with a width of 10 μm at the center of the white dotted circle in this near-field image. At an injection current of 50 mA and above, the optical radiation pattern was concentrated inside the ridge waveguide, and the optical power increased. This concentration indicates that the directivity of the optical radiation pattern was high due to the laser oscillation. At an injection current below 50 mA, on the other hand, the directivity was low; this is because the main components of the optical radiation pattern are spontaneous emission and amplified spontaneous emission (ASE).

The left side of Fig. 1.24b shows images of the TM- and TE-polarization components of the near-field image in Fig. 1.24a. The right side shows the cross-sectional light intensity profiles taken along the white dotted lines in these images. These figures show that the TM-polarization component from the Si laser was spread over the entire device, whereas the TE-polarization component was concentrated at the location of the ridge waveguide. Because the output beam from a conventional semiconductor laser during oscillation contains the TE-polarization [48], this measurement result also confirms that laser oscillation occurred. In this measurement, the intensity ratio of the TE-polarization and the TM-polarization was 8:1. On the other hand, in a conventional semiconductor laser, this intensity ratio is 100:1 or greater [48]. The reason for the difference between the present result and the conventional value is that there is no optical confinement structure in the thickness direction of the ridge waveguide, and also because the spectral wavelength bands of the spontaneous emission and ASE are extremely wide.

Curves A–D in Fig. 1.25a show the output spectral profiles of the Si laser with injection currents of 20, 33, 53, and 60 mA, respectively. In these measurements, another device was used in order to separately observe each longitudinal mode,

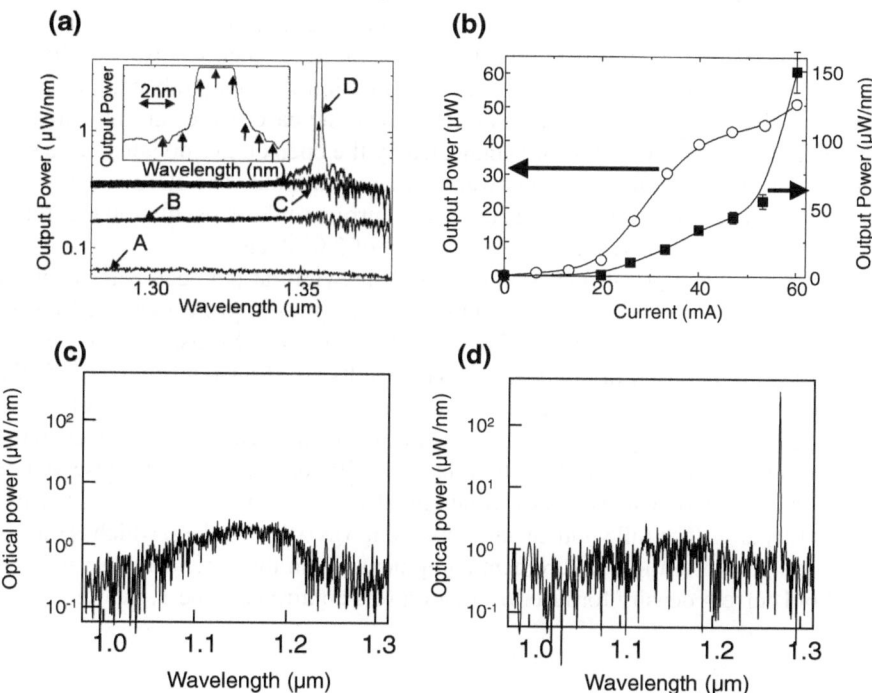

Fig. 1.25 Spectral properties of the Si laser. **a** The output light spectral profiles of the Si laser with a short cavity (cavity length: 250 μm). *Curves A–D* are for injection currents of 20, 33, 53, and 60 mA, respectively. **b** The current dependency of the output optical power. The *open circles* are values obtained by integrating the optical power with respect to wavelength in the wavelength range 1.220–1.380 μm. The *closed squares* are values of the spectral power density at a wavelength of 1.356 μm. **c**, **d** The output light spectral profiles of the Si laser with the long cavity (cavity length: 550 μm) below and above the threshold of current, respectively

which had a shorter cavity (cavity length: 250 μm) than that in Fig. 1.24. A sharp peak was observed at a wavelength of 1.356 μm (the position indicated by the upward arrow on curve D), which shows laser oscillation at an injection current of 60 mA; the tip of the peak went off the top edge of the graph due to saturation of the sensitivity of the photodetector in the measurement equipment. This sharp peak is evidence of photon breeding, as was described in the previous sections. The inset shows a magnified view of the vicinity of the oscillation spectral peak (curve D) during laser oscillation, where several longitudinal modes (positions indicated by the upward arrows in the inset) were observed at a wavelength spacing of 1.0 nm, corresponding to the cavity length (250 μm).

Figure 1.25b shows the current dependency of the output optical power. The open circles are values obtained by integrating the optical power with respect to wavelength in the range 1.220–1.380 μm. At an injection current of 60 mA, the optical output power was 50 μW, and the external differential quantum efficiency was 1 %. This value is as high as that reported for a conventional 1.3 μm-wavelength double heterojunction laser using InGaAsP/InP, which is a direct transition-type semiconductor [49]. The closed squares are values of the spectral power density at a wavelength of 1.356 μm. At the injection current of 60 mA, the spectral power density was 150 μW/nm. Because this device had a wide emission wavelength band, the optical powers of the spontaneous emission and ASE also increase as the current increases. Therefore, the measurement results indicated by the open circles do not show a sudden increase in the optical output power of laser oscillation. The closed squares, on the other hand, do show this, and the threshold current for laser oscillation was found to be 50 mA, giving a threshold current density of 2.0 kA/cm^2.

To measure the optical power and spectral linewidth at the peak wavelength in the laser oscillation spectrum, the output spectral profiles were measured during laser oscillation (at a current of 57 mA) using a different Si laser (cavity length: 500 μm) from that used in Fig. 1.25a, b. The results are shown in Fig. 1.25c, d. The vertical axis of this graph is a logarithmic scale. At an injection current of 55 mA or less (Fig. 1.25c), only a wide emission spectrum was measured. However, above the threshold current of 56 mA, a sharp laser oscillation spectrum was observed, as shown in Fig. 1.25d, with a center wavelength of 1.271 μm, which is also evidence of photon breeding. The full width at half maximum was 0.9 nm or less, which is limited by the resolution of the measurement equipment. From these measured results, the threshold current density for laser oscillation was confirmed to be 1.1 kA/cm^2.

1.9 Light Emitting Diodes Fabricated from Other Crystals

This section reviews the fabrication of LEDs using semiconductor materials other than Si. One is silicon carbide (SiC), which has been considered to be a suitable material for use in high-electrical-power devices because of its high electrical breakdown strength, thermal conductivity, and heat tolerance [50]. However, since SiC is an indirect-transition-type semiconductor, it has never been used for fabricating highly efficient practical LEDs. Another is zinc oxide (ZnO). Although it is a direct-transition-type semiconductor, it has been difficult to form a p-type crystal for practical LEDs.

1.9.1 Using a Silicon Carbide Crystal

Although SiC is an indirect-transition-type semiconductor, research has been conducted for realizing blue LEDs using this crystal. However, the external quantum efficiency of the light emission is low, at 0.1 % or less, and it has not been easy to fabricate practical devices [51, 52]. This section reviews the use of a DPP-assisted process in bulk crystal SiC of the 4H-type (4H-SiC) having a p–n homojunction structure to modify the spatial distribution of the Al dopant for fabricating a blue LED with high light-emission efficiency [53].

Since the bandgap energy E_g of SiC corresponds to ultraviolet to blue wavelengths, visible to ultraviolet SiC-LEDs fabricated via the DPP-assisted process can avoid a decrease in efficiency due to light absorption by 4H-SiC. To achieve light emission from 4H-SiC, the two-step de-excitation process was used, as was described in Sect. 1.3.2. A 500 nm-thick n-type buffer layer (dopant density 1×10^{18} cm^{-3}) was deposited on an n-type 4H-SiC crystal with a thickness of 360 μm, a diameter of 100 mm, a resistivity of 25 m Ω cm, and a surface orientation of (0001), after which a 10 μm-thick n-type epilayer (dopant density 1×10^{16} cm^{-3}) was deposited. The 4H-SiC crystal was then implanted with a p-type dopant (Al) by ion implantation. During this process, the implantation energy was changed in multiple steps in the range 30–700 keV, and the ion dose was adjusted in the range 3.0×10^{13}–2.5×10^{14} cm^{-2} for each implantation energy. As a result, the dopant density was modulated between 2.2×10^{19} and 1.8×10^{19} cm^{-3} in seven periods in the depth direction. After this, thermal annealing was performed for 5 min at 1800 °C to activate the Al ions, forming a p–n homojunction. An ITO film and a Cr/Pt/Au film were deposited on the front and back surfaces, respectively, for use as electrodes. After this, the 4H-SiC crystal was diced to form a device with an area of 500 μm × 500 μm.

A forward bias voltage of 12 V (current density 45 A/cm^2) was applied to the device to bring about annealing due to Joule-heat, which caused the Al to diffuse, modifying the spatial distribution of the dopant concentration. During this process, the device was irradiated from the ITO electrode side with laser light (optical power density 2 W/cm^2) having a photon energy $h\nu_{anneal}$ (=2.33 eV, 532 nm wavelength) smaller than E_g of the 4H-SiC (=3.26 eV) [54]. This induced the DPP-assisted process, which modified the Al diffusion due to annealing, leading to the self-organized formation of unique minute inhomogeneous domain boundaries of Al (Sect. 1.4.1).

Curves A–G in Fig. 1.26a shows the measured light emission spectra from the SiC-LED acquired during DPP-assisted annealing. They were measured at 0.01, 0.25, 0.5, 1.0, 3.0, 8.0, and 24.0 h, respectively, after starting the DPP-assisted annealing.

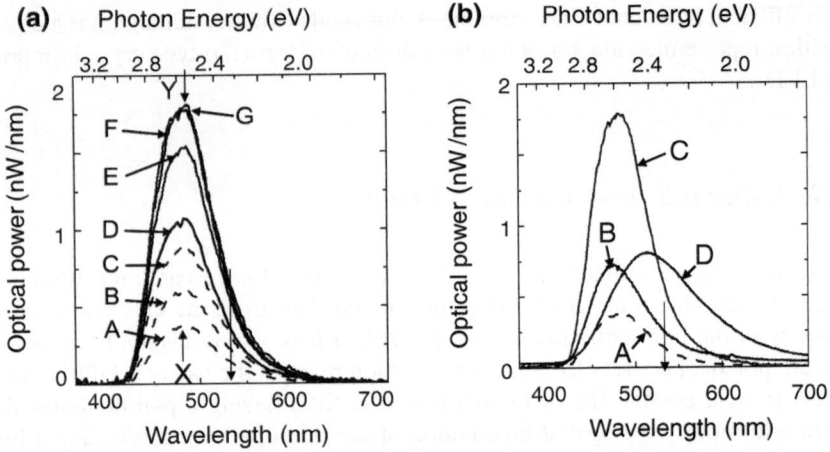

Fig. 1.26 Light emission spectra of the SiC-LED. **a** The spectra acquired during DPP-assisted annealing. *Curves A–G* were acquired at 0.01, 0.25, 0.5, 1.0, 3.0, 8.0, and 24.0 h, respectively, after starting the DPP-assisted annealing. **b** Relationship between the light emission spectral profiles and irradiation intensity during DPP-assisted annealing. The *curve A* was acquired immediately after starting annealing. The *curve B* is the result of annealing without irradiating light. The *curves C* and *D* are the results of annealing with irradiation intensities of 2 and 10 W/cm^2, respectively

The surface temperature during annealing was 120–150 °C. In order to demonstrate photon breeding, the wavelength (the photon energy $h\nu_{anneal}$) of the light irradiated onto the device during annealing is indicated by the long downward arrow in the figure. These curves show that the total emission power (the area surrounded by the curves and horizontal axis in Fig. 1.26a) increases approximately in proportion to the logarithm of the annealing time because the dopant diffusion rate due to annealing has an extremely large distribution, as has been described for two-level systems [55]. After 8 h of annealing (curve F in Fig. 1.26a), the total emission power was five-times greater than the value immediately after starting annealing (curve A). The total emission power saturated when annealing was conducted for 8 h or more. As shown in Fig. 1.26a, the peak emission wavelength was red-shifted from 480 nm (upward arrow X) to 490 nm (downward arrow Y). Curves A–D in Fig. 1.26b shows light emission spectra observed after 8 h of annealing. As shown by the curve D, when the irradiation intensity during annealing was 10 W/cm^2, the wavelength had red-shifted to 515 nm after annealing. It is also shown that the total emission power of the curve D was reduced to about half that in the case of the curve C (irradiation intensity of 2 W/cm^2 during annealing). This was because the progress of the DPP-assisted annealing was not sufficiently controlled by the light irradiation since the 4H-SiC crystal absorbed 532 nm-wavelength light for generating heat. Compared with the curve C, the greater red-shift and also the lower total emission power in the case of the curve D was because the intrinsic energy of the DPP level was close to the photon energy $h\nu_{anneal}$ of the light irradiated during annealing, which counteracted the heat generation due to absorption. On the other hand, when annealing was

Fig. 1.27 Injection current dependency of the light emission power. The SiC-LED was driven with both direct current (*closed circles*) and pulsed current (*closed squares*). The *insets* show the images of output optical beams from the SiC-LED

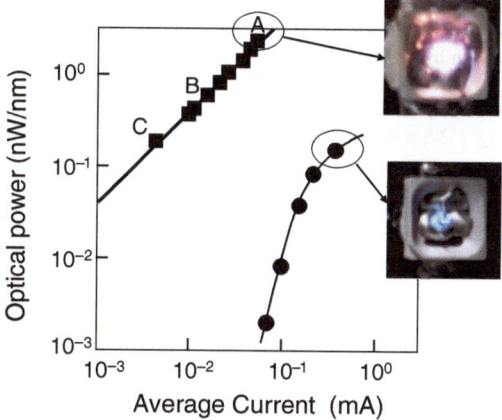

performed using only current, without irradiating any light (curve B in Fig. 1.26b; current density 45 A/cm^2), the electrodes were degraded due to heat generation in about 1 h, and the emission power decreased.

Figure 1.27 shows the measured relations between the injection current and the light emission power. The SiC-LED was driven with both direct current (closed circles) and pulsed current (closed squares). The images of the output optical beams from the SiC-LED shown in the insets were taken under fluorescent room lights. From the measured relationship between the forward bias voltage and the injection current, it was found that a Schottky barrier effect at the electrodes reduced the light emission efficiency. However, this can be eliminated by applying a voltage sufficiently higher than the potential gap of the Schottky barrier. To achieve this, first, the injected direct current was increased, which caused the voltage applied to the device to rise. In this case, as shown by the closed circles and the solid curve in the figure, the emission power increased nonlinearly due to the two-step de-excitation, and at injection currents of 0.3 mA or higher, it saturated. The major cause of this saturation is probably heat generation due to the injection current. Therefore, to avoid this heat generation, second, a pulsed current was injected with a pulse width of 50 μs and a repetition frequency of 100 Hz. In this case, an instantaneous current of 780–1,300 mA flowed in the SiC-LED, and the instantaneous voltage was 23.0–23.6 V. As a result, the emission power was observed to increase linearly and did not saturate, as shown by the closed squares and the solid line in the figure. The reasons for this linear increase are as follows: (1) The driving voltage was sufficiently high, making the number of carriers, which nonradiatively recombine at the Schottky barrier at the electrode, negligibly small. (2) For the injected pulsed current density as high as 320–520 A/cm^2, the Al concentration was not high enough to cause efficient two-step de-excitation. As a result, electrons and holes accumulated at the bottom of the conduction band and the top of the valence band, respectively, where they formed DPPs. Therefore, the one-step transition due to phonon scattering was dominant.

Fig. 1.28 Light emission spectra of the SiC-LED, driven by pulsed current. **a–c** Correspond to squares A, B, and C on the *solid line* in Fig. 1.27, respectively

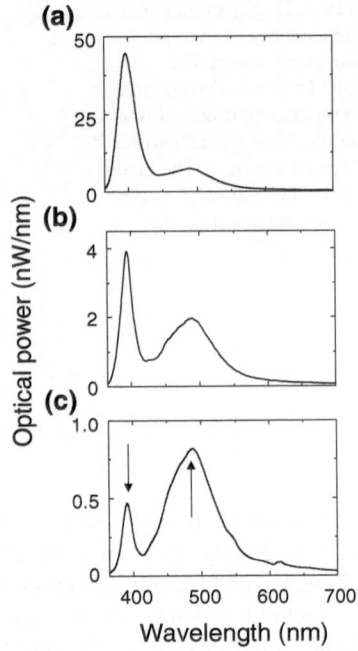

Reason (1) above is well-known in electronic devices having a normal Schottky barrier. The basis for reason (2) is described as follows: Fig. 1.28a–c shows measurement results of the emission spectra of the SiC-LED, driven by pulsed current, which correspond to closed squares A, B, and C on the solid line in Fig. 1.27, respectively. When the pulsed current was small (Fig. 1.28c), the temperature of the device was low, and a separate phonon level was observed (the peak at the wavelength of 390 nm in the figure: downward arrow). This peak appeared for the first time due to pulsed current operation, but the photon energy of this peak is lower than E_g of 4H-SiC by an amount corresponding to the energy of LO-mode phonons (95 meV) or the energy of TO-mode phonons (110 meV) [56]. From a comparison of Fig. 1.28a–c, it is clear that the intensity of this emission peak governs the emission intensity while driving the device with pulsed current. Since the energy of this emission level is close to E_g, recombination luminescence is possible via a one-step transition involving only phonon emission. Of course, an emission peak corresponding to the DPP level also exists (upward arrow in Fig. 1.28c) even under direct-current injection to the SiC-LED. However, comparing Fig. 1.28a–c, the height of the emission peak is saturated. Therefore, when the SiC-LED was driven by a pulsed current, it is likely that the one-step transition was dominant, resulting in a linear increase in the emission power with increasing current.

The gradient of the solid line in Fig. 1.27 corresponds to an external quantum efficiency of 1 %. From the light extraction efficiency (<30 %) and the light absorption (>70 %), the internal quantum efficiency was estimated to be 10 %. This value is as large as the efficiency of conventional LEDs using direct transition-type semiconductors. It is expected that this efficiency will be increased by carrying out more precise adjustments of the thickness and dopant density.

1.9.2 Using a Zinc Oxide Crystal

ZnO is a direct transition-type, wide bandgap semiconductor and is expected to be used as a material for fabricating optical devices such as UV LEDs and laser diodes [57]. However, it is difficult to form a p-type crystal because the acceptors are compensated due to the numerous oxygen vacancies and interstitial zinc in the ZnO crystal [58]. Therefore, despite the numerous efforts that have been made [59–65], there have been very few reports of electroluminescence at room temperature [63, 64]. Because the radius of N ions is approximately the same as that of O ions, they are promising candidates to serve as p-type dopants in ZnO [57], and N doping by ion implantation has been examined [66]. However, the large number of lattice defects generated in normal ion implantation cannot be removed even with thermal annealing, and therefore, no p-type crystals of sufficient quality for fabricating devices have been obtained [66]. This section reviews a p–n homojunction-structured LED that emits visible light at room temperature by applying DPP-assisted annealing to a direct transition-type bulk ZnO crystal [67].

An n-type ZnO single crystal was used, which was grown by the hydrothermal growth method [68]. The crystal axis direction was (0001), the thickness was $500 \, \mu m$, and the electrical resistivity was 50–$150 \, \Omega \, cm$. N^{2+} ions were implanted into the crystal at an energy of $600 \, keV$ and an ion dose density of $1.0 \times 10^{15} \, cm^{-2}$. The implantation depth was confirmed to be about $3 \, \mu m$ by secondary ion mass spectrometry. This allowed the N dopant to be distributed in the vicinity of the crystal surface, forming a p-type ZnO layer. As a result, a p–n homojunction structure was realized. An ITO film was deposited on the surface of the p-type ZnO layer, and a Cr/Al film was deposited on the surface of the n-type ZnO layer to serve as electrodes. Then, the crystal was diced to form a device.

As described in previous sections, when a bulk Si crystal was implanted with B serving as a p-type dopant, the B was readily activated to form acceptors, thus creating a p–n homojunction [19]. In the present case, however, the N dopant was not readily activated [58]. Therefore, first, Joule-heat caused by a forward-bias current was used to anneal the crystal for activating the N dopant. An overview of this principle is as follows: If a forward-bias current is applied directly after implanting the N dopant, the N is not activated sufficiently. Therefore, only electrons, which are majority carriers, carry the electrical current. Also, because no holes exist, recombination emission does not take place either. Therefore, the p–n homojunction remains highly resistive, and when a constant current is applied, the voltage applied across the p–n junction is high. As a result, a high level of Joule-heat occurs, diffusing the N dopant and considerably changing the concentration distribution, causing the N dopant to be activated. Therefore, since holes also become current carriers, the resistance decreases.

Because ZnO is a direct-bandgap semiconductor, the electrical energy is converted to spontaneously emitted optical energy through recombination of electrons and holes, and this is radiated from the crystal to the outside. The Joule-heat drops due to this energy dissipation and the decrease in resistance mentioned above, and therefore, the concentration distribution of the N dopant eventually reaches a steady state. This completes the N dopant activation process.

Second, DPP-assisted annealing was used simultaneously with the activation described above. The process can be explained as follows: A p–n homojunction is formed in the ZnO crystal by implanting N^{2+} ions. However, because this structure is simple, the electrons and holes both exhibit wide spatial distributions. Therefore, their recombination probability is low, and the emission intensity is also low. The DPP-assisted process is used to increase the emission intensity. Specifically, while applying a forward bias current during the annealing, the crystal surface is irradiated with light having a small photon energy $h\nu_{anneal}$ compared with E_g of ZnO. The mechanism of the DPP-assisted annealing of the ZnO crystal is equivalent to that described in Sect. 1.5.1. As a result, an N dopant concentration distribution that is suitable for inducing the DPP-assisted process with high efficiency is formed in a self-organized manner.

The following two types of devices were fabricated.

Device 1 A device fabricated by annealing with a forward-bias current alone, without light irradiation, to activate the N dopant.

Device 2 A device fabricated by DPP-assisted annealing. To do so, during annealing with the forward-bias current, the device was irradiated with laser light having a photon energy $h\nu_{anneal}$ (=3.05 eV, 407 nm wavelength), which is smaller than E_g (=3.4 eV) of ZnO.

The forward-bias current density for annealing both devices was $0.22\,A/cm^2$. For Device 2, the irradiation power density was $2.2\,W/cm^2$. Figure 1.29a shows the change in surface temperature with time at the center of Device 2. The surface temperature rose to $100\,°C$ when annealing commenced, and then dropped, reaching a constant temperature of $74\,°C$ after about 60 min. This temperature drop was caused by the generation of DPPs as annealing progressed, bringing about stimulated emission, and by part of the electrical energy added to produce Joule-heat being dissipated in the form of optical energy. To confirm this stimulated emission effect, Fig. 1.29b shows the results of measuring the change in surface temperature with time for another identical sample of Device 2, when the irradiated light power was turned on and off every 5 min. The temperature dropped during light irradiation, confirming the stimulated emission effect.

As for the device operation, first, the curves A–C in Fig. 1.30 are light emission spectra of Device 1 at room temperature with forward-bias currents of 10, 15, and 20 mA, respectively. These spectra are composed of a high-intensity, narrow-band emission component in the ultraviolet region close to a wavelength of 382 nm and a low-intensity, wide-band emission component in the visible region above 490 nm. The former is attributed to the band edge transition in ZnO, and the latter is attributed to emission from defect levels [69]. The emission from the defect levels is not related

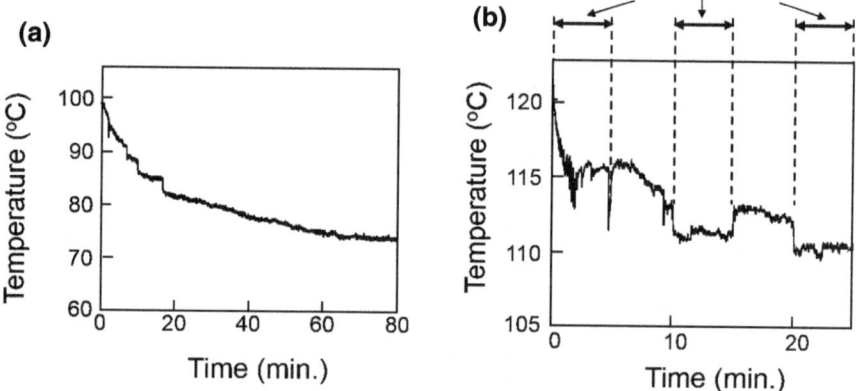

Fig. 1.29 Change in surface temperature with time at the center of Device 2. **a** The change in the surface temperature in the process of annealing with light irradiation. **b** The change in surface temperature for another identical sample of Device 2, when the irradiated light power was turned on and off every 5 min

Fig. 1.30 Light emission spectra of Device 1. The *curves A, B,* and *C* show the results obtained with forward bias currents of 10, 15, and 20 mA, respectively

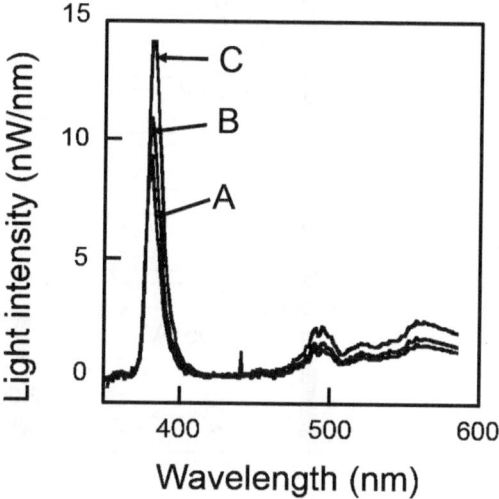

to the DPP-assisted process in the case of Device 2, described below, and depends on the crystal quality. In other words, from these spectra, the emission from Device 1 was confirmed to be mainly due to the band edge transition. The $V - I$ characteristic of this device showed the same rectifying properties as an ordinary diode, confirming that a suitable p–n homojunction was formed by the annealing.

Next, light emission spectra of Device 2 at room temperature are shown by the three curves A, B, and C in Fig. 1.31a. Curves A, B, and C shows the results obtained with forward-bias currents of 10, 15, and 20 mA, respectively. The emission peak wavelength of curve A was 393 nm, which is attributed to the band edge transition,

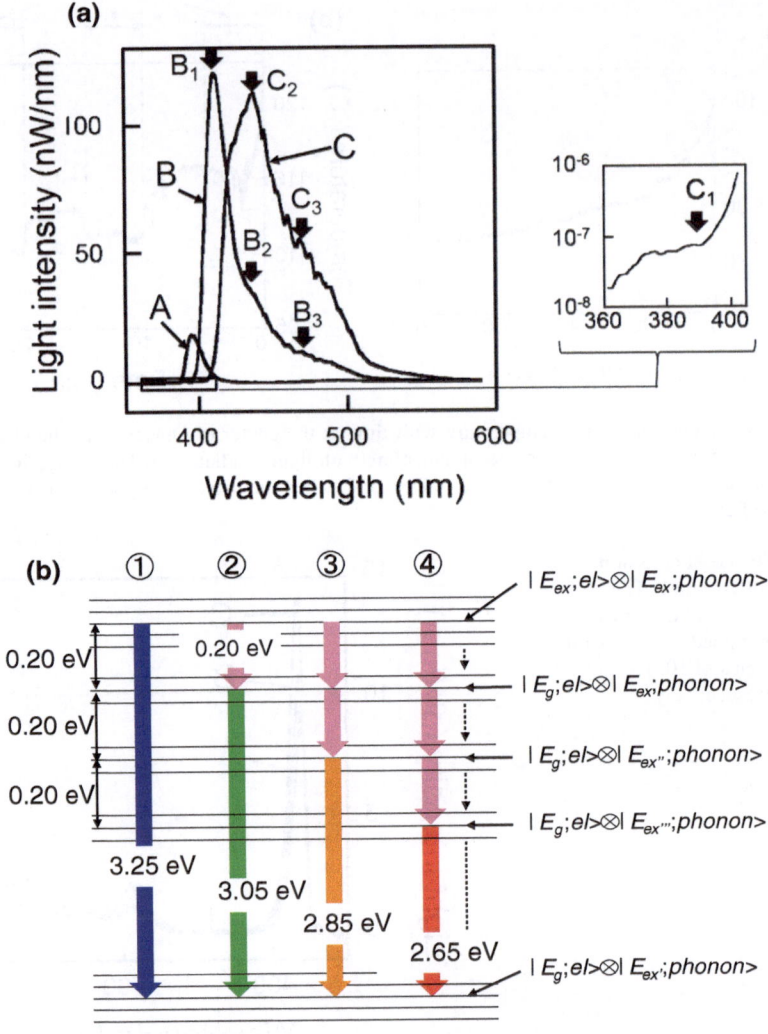

Fig. 1.31 Light emission characteristics of Device 2. **a** Light emission spectra at room temperature. The *curves A, B,* and *C* show the spectra measured with forward bias currents of 10, 15, and 20 mA, respectively. **b** Schematic explanation of the two-step light emission during LED operation

similar to the case of Device 1 (Fig. 1.30). Comparing curves B and C with curve A, the spectral centroid of the emission spectrum exhibits a red-shift as the forward-bias current was increased. However, this shift was not attributed to a change in E_g induced by Joule-heat [70], because the emission peak attributed to the band edge transition in Device 1 did not show this kind of shift. Furthermore, as shown in the inset of this figure, the weak emission (arrow C_1), attributed to the band edge transition, was also found in curve C. Therefore, the red-shift of the emission spectra, which is attributed to Joule-heat at forward-bias currents in the range of 10–20 mA, can be neglected. The reason why this red-shift is attributed to stimulated emission driven by DPPs, not Joule-heat, is as follows.

Light emission during LED operation, like the stimulated emission process during annealing, is attributed to a two-step transition, as was described in Sect. 1.3.2. This transition is illustrated in Fig. 1.31b in the present case, i.e., from $|E_{ex}; el\rangle \otimes |E_{ex}; phonon\rangle$ to $|E_g; el\rangle \otimes |E_{ex}; phonon\rangle$ (the pink arrow at ② in Fig. 1.31b) and from $|E_g; el\rangle \otimes |E_{ex}; phonon\rangle$ to $|E_g; el\rangle \otimes |E_{ex'}; phonon\rangle$ (the green arrow at ② in Fig. 1.31b). Because the second-step transition is an electric dipole-forbidden transition, only DPPs are emitted, and these DPPs are scattered by the inhomogeneously distributed N dopant and are converted to propagating light. The photon energy of the emitted light is determined by the photon energy $h\nu_{anneal}$ of the light radiated during annealing, i.e., photon breeding takes place. This is because the N dopant concentration distribution is formed in a self-organized manner by the DPP-assisted annealing, with the result that a transition via the intermediate phonon level corresponding to $h\nu_{anneal}$ easily occurs in Device 2. Therefore, in curve B in Fig. 1.31a, the peak photon energy of 3.03 eV (409 nm wavelength, arrow B_1) is almost equal to $h\nu_{anneal}$.

On the other hand, the photon energy of light generated in the first-step transition is given by the energy difference between $|E_{ex}; el\rangle \otimes |E_{ex}; phonon\rangle$ and $|E_g; el\rangle \otimes |E_{ex}; phonon\rangle$. Comparing the blue arrow at ① and the green arrow at ② in Fig. 1.31b, this difference is 0.20 eV. (The blue arrow at ① represents the band edge transition of ZnO, whose energy is 3.25 eV (382 nm wavelength) according to Fig. 1.30.) This first-step transition is an electric dipole-allowed transition. However, because the occupation probability of such high-energy phonons is low, propagating light is not emitted, and only DPPs are generated. In addition, when the stimulated emission driven by the DPPs emitted in the first-step transition is repeated one or two more times, new intermediate phonon levels $|E_g; el\rangle \otimes |E_{ex''}; phonon\rangle$ and $|E_g; el\rangle \otimes |E_{ex'''}; phonon\rangle$ whose energies are lower by amounts corresponding to the energy of the DPPs are formed. Thus, the energies of the photons emitted via these new intermediate phonon levels are lower by amounts corresponding to the energy of the DPPs emitted in the first-step transition (0.20 eV). In the case of ③ and ④ in Fig. 1.31b, these energies are 2.85 eV (435 nm) and 2.65 eV (468 nm), respectively, as shown by the yellow arrow and the red arrow. These are similar to the emission peaks measured in the conventional optical transition of bulk ZnO crystal, which are an integer multiple of the LO-mode phonon energy (72 meV) [69]. However, unlike the conventional electric dipole-allowed transition, in the transition driven by DPPs, the exchanged energy is an integer multiple of the energy determined by DPPs among the multiple phonon modes involved, rather than a material-specific phonon mode. The slope of curve B in Fig. 1.31a shows two bumps (arrows B_2 and B_3), whose positions were found to be 2.84 eV (436 nm) and 2.64 eV (470 nm), respectively, by the curve fitting based on the second-derivative spectroscopy method [71]. These agree well with the photon energies indicated by the yellow and red arrows at ③ and ④ in Fig. 1.31b. In addition, curve C also shows a peak (arrow C_2) and one bump (arrow C_3) at 2.84 eV (436 nm) and 2.63 eV (471 nm), respectively. These also

Fig. 1.32 The dependency of the output optical power of the two devices on the forward bias current. The *curves A* and *B* are for Devices 1 and 2, respectively

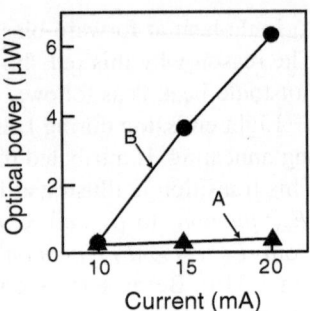

agree well with the photon energies indicated by the yellow and red arrows at ③ and ④ in Fig. 1.31b.

Light emission via the intermediate phonon levels occurs not only due to two-step stimulated emission of ③ and ④ in Fig. 1.31b but also due to stimulated emission involving three or more steps. Therefore, when the forward bias current is increased, the emission intensity at the low-energy side also increases, which explains the redshift in Fig. 1.31a. Note that the emission intensity due to the interband transition in curve B is small; this is attributed to the fact that the transition ② involving stimulated emission is faster than the transition ① involving only spontaneous emission, because electrons in the conduction band for stimulated emission relax to the intermediate phonon level.

To compare the performance of Devices 1 and 2, curves A and B in Fig. 1.32 respectively show the dependency of the output optical power of the two devices on the forward-bias current. These output optical powers were obtained by integrating the curves in Figs. 1.30 and 1.31a in the wavelength range 350–600 nm. For curve B, at the forward bias current of 20 mA (current density 0.22 A/cm^2), the optical output power from Device 2 was 6.2 μW, which was about 15-times higher than that from Device 1, shown in curve A.

1.10 Other Devices

This section reviews applications in which DPs and DPPs can be used not only for LEDs and lasers but also for other optical devices. The first example is an optical and electrical relaxation oscillator, and the second one is an infrared photodetector with optical amplification.

1.10.1 Optical and Electrical Relaxation Oscillator

Optical pulse oscillators have been widely used in the fields of optical communication, optical data storage, optical fabrication, spectroscopy, and so on. Mode-locked lasers and semiconductor lasers driven by pulsed current are popular examples of such devices [72]. The former are large in size and have high power consumption because they consist of numerous electronic parts and optical elements. Although the latter are compact, they need to be driven by an electrical trigger and thus require complicated electrical driving circuits.

This section reviews a novel optical and electrical relaxation oscillator by using the Si-LED described in Sect. 1.5 [73]. Because the device is operated by connecting it only to a DC power supply, and no optical elements are required, it is expected to solve the problems described above. For this purpose, two Si-LEDs were fabricated by the DPP-assisted annealing described in Sect. 1.5.1. They are:

Si-LED1 Fabricated by applying a voltage of about 10 V and an injection current of about 420 mA for 30 min under laser light irradiation (power: 500 mW)

Si-LED2 Fabricated by applying a voltage of about 7.2 V and an injection current of about 700 mA for 30 min under laser light irradiation (power: 200 mW)

Si-LED1 was mainly used for simulation and measurement of the spontaneous emission lifetime, whereas Si-LED2 was used for quantitative characterization of the oscillation. These Si-LEDs emit infrared light whose emission spectrum has a peak that corresponds to the photon energy $h\nu_{anneal}$ (=0.95 eV, 1.30 μm wavelength) of the irradiated laser light, which is evidence of photon breeding. In addition, because of the considerably inhomogeneous spatial distribution of the B concentration in the Si-LED, the current density also becomes inhomogeneous when the device has a size as large as 9 mm^2 in area and 650 μm in thickness. As a result, the $V - I$ characteristic curve is S-shaped, showing a breakover voltage V_b, as was the case with Fig. 1.12a. In the following, the S-shaped characteristic is expressed as $I = f(V)$. This S-shaped characteristic, i.e., a negative resistance characteristic, was utilized to realize an optical and electrical relaxation oscillator.

The curves A and B in Fig. 1.33 shows the measured results of the temporal profiles of the optical power and the voltage of the Si-LED2, acquired at room temperature. Comparing the two curves, they varied synchronously, and the oscillation frequency was 10 kHz. The amplitudes of the curves A and B were 2.5 mW$_{p-p}$ and 47 V$_{p-p}$, respectively. The optical energy integrated over one period of oscillation was 0.14 μJ.

Fig. 1.33 The measured temporal profiles of the output signals from the Si-LED2. The *curves A* and *B* are the optical power and the voltage, respectively

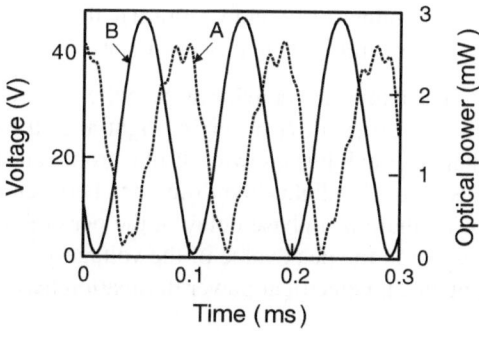

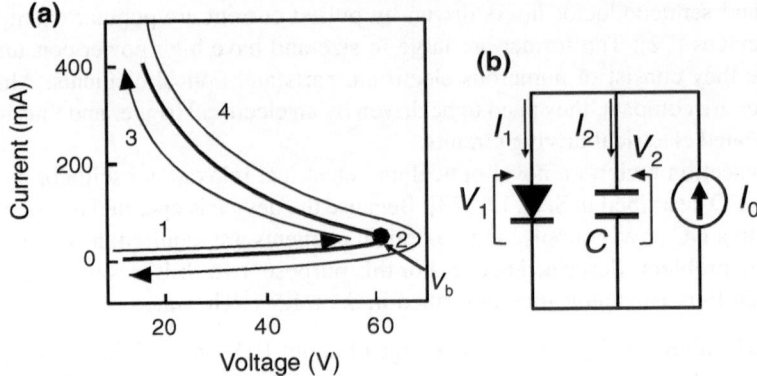

Fig. 1.34 The S-shaped $V - I$ characteristic (**a**) and an equivalent electrical circuit (**b**)

The S-shaped $V - I$ characteristic is shown in Fig. 1.34a. The Si-LED can be treated as an electrical circuit in which the Si-LED is connected in parallel with a capacitor and a constant-current source, as shown in Fig. 1.34b, where the capacitor can be regarded as the stray capacitance of the Si-LED and/or the circuit wiring. Here, C is the capacitance of the capacitor, I_1 and I_2 are the currents that flow through the Si-LED and the capacitor, respectively, V_1 and V_2 are the voltages applied to them, and I_0 is the current from the constant-current source. By setting $I_0 > f(V_b)$ and $V_1 = V_2 = 0$ as the initial conditions, it is expected that the optical power and voltage will exhibit periodic and oscillatory temporal behaviors due to sequential processes 1–4 described below and schematically explained in Fig. 1.34a.

1. V_1 increases as electric charges flow into C, and the current flows into the Si-LED simultaneously.
2. At the moment V_1 reaches V_b, V_2 increases because the current $I_0 - f(V_b)$ flows into C. However, because V_1 cannot exceed V_b, the difference between V_1 and V_2 increases.
3. I_1 increases and, as a result, V_1 decreases because the electric charge is released from C due to the voltage difference between V_1 and V_2. This accelerates the release of the charge and, as a result, the optical output power increases rapidly.
4. The values of V_1 and V_2 return to the initial state due to the decrease of the electric charge in C. Then, process 1 starts again.

The temporal behavior of the optical power exhibits a pulse-like profile due to the rapid increase of I_1 occurring right after the electric discharge. On the other hand, that of V_1 exhibits a sawtooth-like profile due to the instantaneous electric discharge. Because of a slight time difference between the electric discharge and the increase of I_1, there is a phase delay in the periodic pulse profile of the optical power with respect to the peak value of the voltage V_1. Since these temporal profiles are due to optical and electrical power dissipation based on the difference between V_1 and V_2,

Fig. 1.35 Simplified
equivalent circuit of a Si-LED

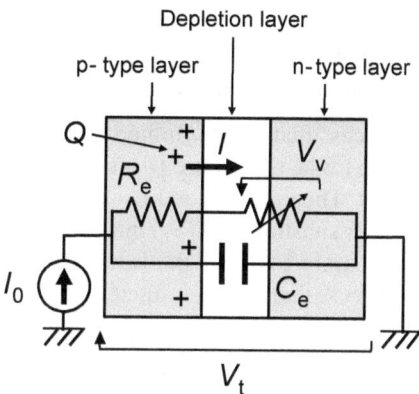

this circuit is regarded as a nonequilibrium open system. Thus, the temporal behavior of the voltage and optical power described above is called relaxation oscillation [20].

Figure 1.35 shows a simplified equivalent circuit of a Si-LED [74, 75], which is composed of a p-type layer, an n-type layer, and a depletion layer. For simplicity, carriers are assumed to recombine only in the depletion layer. C_e is the equivalent capacitance representing the capability of storing electric charge Q in the Si-LED, R_e is the resistance at the p-type layer, I_0 is the current from the constant-current source, I is the current injected into the depletion layer, V_v is the difference in the voltage between the anode-side and the cathode-side of the depletion layer, and V_t is the total voltage applied to the Si-LED.

The temporal variations in the number of carriers n in the Si-LED and the number of photons p in the depletion layer are represented by the following rate equations [76]:

$$\frac{dn}{dt} = \frac{f(V)}{q} - \frac{n}{\tau} - G(n - n_{th})p \qquad (1.1)$$

and

$$\frac{dp}{dt} = \frac{n}{\tau} + BG(n - n_{th})p - Ap \qquad (1.2)$$

The first, second, and third terms on the right side of (1.1) represent electron injection, spontaneous emission, and stimulated emission, respectively. Those in (1.2) represent spontaneous emission, stimulated emission, and light emission from the Si-LED, respectively. Here, e is the electron charge, τ is the spontaneous emission lifetime, G is the stimulated emission coefficient, n_{th} is the transparency carrier number in the depletion layer, B is the coefficient of electron confinement, and A is the rate of photon dissipation from the depletion layer to the outside. For simplicity, the carrier injection efficiency is assumed to be unity.

The simulation was performed using (1.1), (1.2), and circuit equations for the equivalent circuit shown in Fig. 1.35. The values of n and p were derived iteratively with the time increment dt by the following steps:

1. Process that is rate-limited by charging of C_e: In the processes 1 and 2 shown in Fig. 1.34a, the charge Q at time $t + dt$ was derived using $Q(t+dt) = Q(t) + (I_0 - I)dt$. Then, V_t, V_v, and I were derived by $V_t = Q/C_e$, $V_v = V_t - R_e I$, $I = f(V_t)$. After substituting $f(V_t)$ into (1.1), (1.1) and (1.2) were approximated as difference equations to derive the values of n and p at time $t + dt$.
2. Process that is rate-limited by discharging of C_e:

   2.1 If $V_1 > V_b$ in process 1, the capacitor C_e started discharging. Thus, V_t, V_v, and I were set to $V_t = V_v = V_b$ and $I = I_0$. Then, n and p at time $t + dt$ were derived using (1.1) and (1.2).

   2.2 If the discharging of C_e continued, V_t and V_v were derived using $V_t = f^{-1}(I)$ and $V_v = Q/C_e$. Then, $Q(t + dt)$ and $I(t + dt)$ were derived using $Q(t + dt) = Q(t)(V_t - V_v)dt/R_e$ and $I(t + dt) = I_0(V_t - V_v)/R_e$. This calculation continued until $V_2 > V_b$ or $I < 0$ was satisfied.

The values of physical quantities R_e, C_e, $f(V)$, τ, G, n_{th}, B, and A have to be determined to perform the simulation. Among them, known values were employed for R_e, C_e, n_{th}, B, and A, as will be shown later. The curve in Fig. 1.12a was used for $f(V)$. However, the spontaneous emission lifetime τ must be found through measurements because electroluminescence from an indirect-transition-type semiconductor has never been observed, and hence the value of τ for a bulk Si crystal is unknown.

The Si-LED1 was used for the direct measurement of τ, and the non-annealed Si wafer was also used as a reference specimen. As an excitation light source, the second-harmonic pulsed light from a Ti-sapphire mode-locked laser (2 ps pulse width, 80 MHz pulse repetition rate, 454 nm wavelength, 30 mW power) was used. The light passed through a longpass filter with a cutoff wavelength of 850 nm or 1,000 nm before reaching the Si-LED and the non-annealed Si wafer. The specimens were placed in a vacuum chamber and cooled down to about 6 K.

Curves A and B in Fig. 1.36a represent the acquired spectral profiles of the Si-LED1 and the non-annealed Si wafer, respectively. Neither curve clearly shows band-edge emission at wavelengths around 1.11 μm, which corresponds to E_g of Si. On the contrary, the spectra were broadened and extended to the longer-wavelength region, due to a multi-step de-excitation inherent to the DPP-assisted process. Temporal decreases in the photoluminescence intensity were measured for the Si-LED1 and the non-annealed Si wafer after optical pulse irradiation, as shown by the curves A and B in Fig. 1.36b. An exponential function was used for least squares fitting to the measured values:

$$y(t) = y_0 + y_1 \exp\left(-t/\tau\right) \tag{1.3}$$

Fig. 1.36 Measured
photoluminescence spectral
profiles of the Si-LED (**a**) and
temporal decreases in
intensity (**b**). *Curves A and B*
are for the Si-LED1 and the
non-annealed Si wafer,
respectively

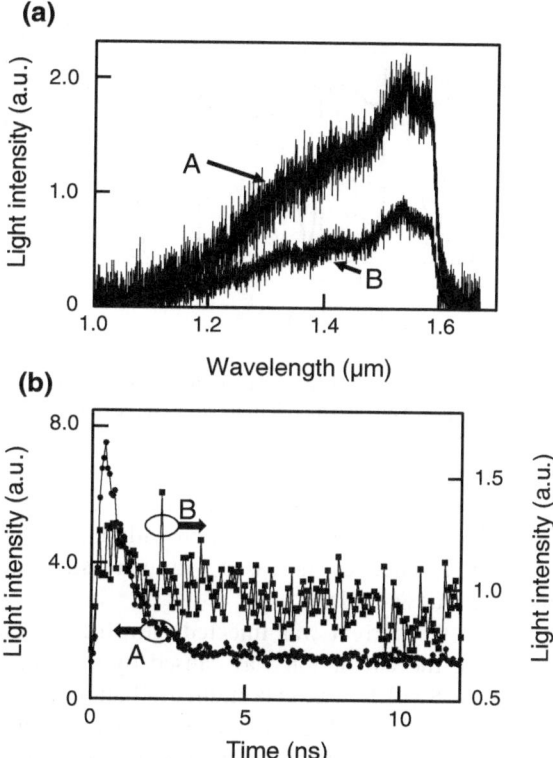

As a result, the value $\tau = 0.90 \pm 0.02$ ns was obtained from the curve A. This value was used for the simulation. In the case of curve B, on the other hand, the spontaneous emission probability was very low since the non-annealed Si wafer remained an indirect transition-type semiconductor. As a result, the values of curve B were very small, which made it difficult to estimate the value of τ. Comparing the curves A and B, the remarkable increase in the spontaneous emission probability enabled the first successful estimation of τ by this method. The value of τ derived above is as short as that of direct transition-type semiconductors [77–79].

The values of the other physical quantities used for the simulations were: $R_e = 5\,\Omega$, $C_e = 1.5 \times 10^{-10}$ F, $n_{th} = 7.0 \times 10^{13}$, and $B = 0.1$. For A, an equation $A = 1 - r^{1/t_0} s^{-1}$ was used, where $r = 0.0002$ is the Fresnel reflection coefficient at the boundary of the depletion layer and the p-type/n-type layer, and $t_0 = 31$ fs is the time for the light to traverse the depletion layer. The value of G was used as a fitting parameter for the simulation.

The results of the simulation are shown by curves A and B in Fig. 1.37a. The photon number and voltage varied synchronously, and the temporal behavior of the photon number (curve A) exhibited a pulse-like profile, whereas that of the voltage (curve B) was sawtooth-like. The pulse width of curve A for the optical power depended

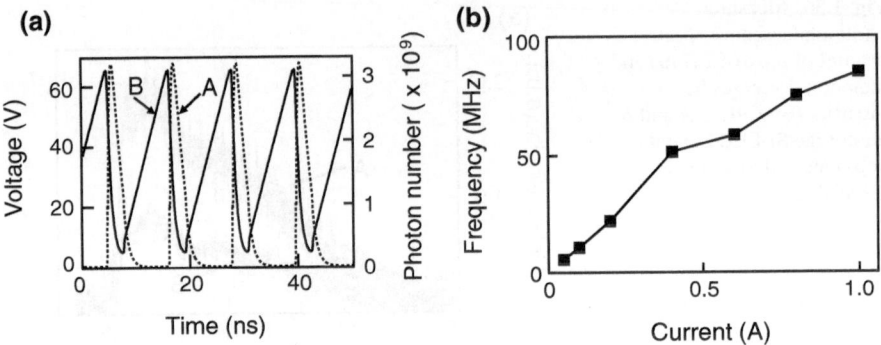

Fig. 1.37 The results of the simulation. **a** The *curves A* and *B* represent the photon number and voltage, respectively. **b** The relation between the injection current and oscillation frequency

on the spontaneous emission lifetime τ. The amplitude of the curve A decreased and exhibited relaxation oscillation with increasing G. As shown by Fig. 1.37b, the oscillation frequency increased with increasing injection current I_0, and the rate of increase was 80 MHz/A.

When the current was injected into Si-LED1 without connecting any external capacitor, the oscillation was caused by stray parasitic capacitance in the circuit and/or the Si-LED. The maximum oscillation frequency was 34 kHz. Comparing the two curves in Fig. 1.37a, the optical power and voltage oscillated synchronously, and the optical power took the maximum value at a time slightly after the voltage took the maximum value. This agrees with the measured result shown in Fig. 1.33. The reason why the two curves in Fig. 1.33 change more slowly than those of Fig. 1.37a is due to stray capacitance, stray inductance, and parasitic resistance of the circuit.

1.10.2 Infrared Photodetector with Optical Amplification

Si photodetectors (Si-PDs) are widely used photoelectric conversion devices. However, the long-wavelength cut-off (1.11 μm) of their photosensitivity is limited by Eg of Si [80]. Because of this, materials such as Ge [81], InGaAsP [82], and InGaAs [83] that have a smaller E_g than Si have been used in optical fiber communications. However, Ge photodetectors have a large dark current, and cooling is required in many cases. In addition, InGaAs photodetectors suffer from problems such as the use of highly toxic metal-organic materials in their fabrication, high cost, and so forth. Recently, depletion of resources, such as In, has also been a problem. If the photosensitivity limit of Si-PDs could be extended into the infrared region at 1.3 μm and above, these problems could be solved. An additional benefit of Si-PDs is their high compatibility with electronic devices. For this reason, photoelectric conversion devices exploiting effects such as mid-bandgap absorption [84–86], surface-state absorption [87, 88], internal photoemission absorption [89, 90], and two-photon absorption [91, 92] in Si have been reported. However, in the case of mid-bandgap absorption, for example, the photosensitivity at a wavelength of $1.3\,\mu m$ is limited to only $50\,mA/W$ [84].

This section reviews a novel Si-PD with increased photosensitivity with optical amplification [93]. Specifically, the DPP-assisted annealing is performed to control the spatial distribution of the B concentration in a Si crystal in a self-organized manner to efficiently generate DPPs. The photocurrent of the novel Si-PD fabricated by this method is varied by a stimulated emission process driven by the incident light. Because this stimulated emission process causes optical amplification, the photosensitivity of this Si-PD remarkably increases.

The operating principle of the fabricated Si-PD is based on a DPP-assisted process. By using DPPs, it is possible to create an electron via a two-step excitation even with photons having an energy smaller than E_g of Si. Therefore, the Si-PD can exhibit photosensitivity even for infrared light with a photon energy smaller than E_g.

Electrons in the Si-PD experience a two-step excitation described below.

First step The electron is excited from the initial ground state $|E_g; el\rangle \otimes |E_{thermal}; phonon\rangle$ to intermediate state $|E_g; el\rangle \otimes |E_{ex}; phonon\rangle$. Here, $|E_g; el\rangle$ represents the ground state (valence band) of the electron, and $|E_{thermal}; phonon\rangle$ and $|E_{ex}; phonon\rangle$ respectively represent the thermal equilibrium state of the phonon determined by the crystal lattice temperature and the excited state of the phonon. Because this is an electric dipole-forbidden transition, a DPP is essential for the excitation.

Second step The electron is excited from the intermediate state $|E_g; el\rangle \otimes |E_{ex}; phonon\rangle$ to the final state $|E_{ex}; el\rangle \otimes |E_{ex}; phonon\rangle$. Here, $|E_{ex}; el\rangle$ represents the excited state (conduction band) of the electron, and $|E_{ex}; phonon\rangle$ represents the excited state of the phonon. Because this is an electric dipole-allowed transition, the electron is excited not only by the DPP but also by propagating light. After this excitation, the phonon in the excited state relaxes to a thermal equilibrium state having an occupation probability determined by the crystal lattice temperature, which completes excitation to the electron excited state $|E_{ex}; el\rangle \otimes |E_{thermal}; phonon\rangle$.

When light having a photon energy smaller than E_g is incident on the Si-PD, electrons are excited by the two-step excitation described above, generating a photocurrent. Photosensitivity to this incident light is manifested by means of the above process. Note that applying a forward current to the Si-PD causes the two-step stimulated emission described in Sect. 1.3.2. Here, if the electron number densities occupying the state $|E_{ex}; el\rangle \otimes |E_{thermal}; phonon\rangle$ and the state $|E_g; el\rangle \otimes |E_{ex}; phonon\rangle$, satisfy the Bernard–Duraffourg inversion condition, the number of photons created by stimulated emission exceeds the number of photons annihilated by absorption. In other words, optical amplification occurs. Because the amplified light brings about the two-step stimulated emission again via DPPs, the photosensitivity of the Si-PD far exceeds the photosensitivity based on only the two-step excitation process.

To realize the optical amplification described above, it is essential to efficiently generate DPPs in the p–n homojunction of the Si-PD. To do so, the fabrication method of Sect. 5.1 was adopted. First, an n-type Si crystal with an electrical resistivity of $10\,\Omega\,cm$ and a thickness of $625\,\mu m$, doped with As, was used. This crystal was doped with B via ion implantation to form a p-type layer. For the B doping, the implantation energy was $700\,keV$, and the ion dose density was $5 \times 10^{13}\,cm^{-2}$. After forming a p–n homojunction in this way, an ITO film and a Cr/Al film were deposited for use as electrodes. Then, the Si crystal was diced to form a device.

Second, DPP-assisted annealing was performed by applying a forward current to the Si-PD to generate Joule-heat, causing the B to diffuse and changing the spatial distribution of the B concentration. During annealing, the device was irradiated, from the ITO electrode side, with laser light having a photon energy $h\nu_{anneal}$ ($0.94\,eV$, $1.32\,\mu m$ wavelength) smaller than E_g of Si.

This method is the same as the method in Sect. 1.5.1. Here, however, in order to make use of the stimulated emission process for the Si-PD to be fabricated, it is necessary to make the probability of stimulated emission larger than the probability of spontaneous emission. To do so, the forward current density for annealing was kept smaller than that of the two-step stimulated emission, namely, $1.3\,A/cm^2$. As a result, the number of injected electrons per unit time and per unit area was determined to be $8.1 \times 10^{18}\,s^{-1}\,cm^{-2}$, which corresponds to the probability of spontaneous emission. On the other hand, the probability of stimulated emission corresponds to the number of photons per unit time and per unit area, which is $3.9 \times 10^{19}\,s^{-1}cm^{-2}$ in the case of the laser power of $120\,mW$ used here. Comparing this with the number of injected electrons confirms that the probability of stimulated emission is sufficiently large. The fabricated Si-PD was evaluated by analyzing its optical and electrical properties, which are described as follows.

(1) The spectral photosensitivity was measured without injecting a forward current to the device. The results are shown by curves A–C in Fig. 1.38. Curve A shows the values obtained with a Si-PD fabricated by the DPP-assisted annealing. For comparison, curve B shows values obtained with a Si-PD fabricated without annealing. Curve C shows the values obtained with a Si-PIN photodiode (Hamamatsu Photonics, S3590) used as a reference. In the wavelength region longer than the cutoff wavelength ($1.11\,\mu m$), curve A shows more gentle reduction of the photosensitivity, and its value is about three-times higher than that of curve C at wavelengths above $1.16\,\mu m$. In addition, the photosensitivity for curve A is larger than that for curve B. This is due to the spatial distribution of the B concentration being controlled in a self-organized manner by the DPP-assisted annealing so that DPPs are efficiently generated. Also, higher photosensitivity for curve B compared with that for curve C indicates that DPPs are readily generated inside the Si-PD compared with the case of curve C, as a result of implantation of a high concentration of B.

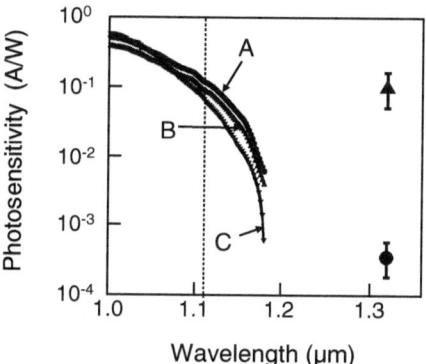

Fig. 1.38 Measured spectral photosensitivities. *Curves A* and *B* show the values obtained with a Si-PD fabricated with and without the DPP-assisted annealing, respectively. *Curve C* shows the values obtained with a Si-PIN photodiode (Hamamatsu Photonics, S3590) used as a reference. The *closed circle* and the *closed triangle* are the measured values for cases where the forward current densities of the fabricated Si–PD are 60 mA/cm² and 9 A/cm², respectively

Because DPP-assisted annealing was performed while radiating 1.32 μm-wavelength light, it is expected that the photosensitivity will be selectively increased when light having the same wavelength (1.32 μm) is incident on the device. This wavelength-selective photosensitivity increase corresponds to photon breeding in the case of the LED and laser in the previous sections. For reference, this increase has already been observed in the case of organic photovoltaic devices that have been developed using the DPP-assisted process [94]. In the following, the photosensitivity of the Si-PD for incident light with a wavelength of 1.32 μm in particular is discussed: A constant forward current was injected into the device, and the photosensitivity was evaluated when the wavelength of the incident light was 1.32 μm. Photoelectric conversion in this case involves not only the two-step excitation, but also the two-step stimulated emission process. Here, the contribution of the latter is sufficiently large. A semiconductor laser was used as the light source, and the output beam was made incident on the Si-PD after being intensity-modulated with a chopper. The current variation $\Delta I = V/R$ was obtained by the measured voltage variation V due to this incident light, where R is the resistance of the Si-PD. Then, it was divided by the incident light power P to obtain the photosensitivity $\Delta I/P$. The results are indicated by the closed circle and the closed triangle in Fig. 1.38. They are the measured values for cases where the forward current densities of the fabricated Si-PD are 60 mA/cm² and 9 A/cm², respectively. The photosensitivity for the current density of 9 A/cm² is 0.10 A/W. This is as much as two-times higher than the case using mid-bandgap absorption described above, demonstrating the increased photosensitivity. This value is about 300-times higher than the 60 mA/cm² case, and is as large as the value of curve C at a wavelength of 1.09 μm. This photosensitivity is sufficiently high for use in long-distance optical fiber communication systems [95].

Fig. 1.39 Measured voltage–current characteristics. The *curves A* and *B* are for cases where the Si–PD was irradiated and not irradiated with light, respectively

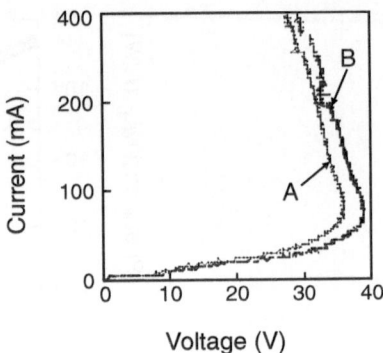

The increase in photosensitivity with increasing forward current at a wavelength of 1.32 μm is due to the higher stimulated emission gain, as well as the higher number of recombining electrons.

(2) The voltage–current characteristics were evaluated for cases where the Si-PD was irradiated and not irradiated with 1.32 μm-wavelength, 120 mW-power laser light. The measurement results are shown by curves A and B in Fig. 1.39. Both curves show negative-resistance characteristics at forward currents of 80 mA and higher, similarly to the case of the large-area infrared Si-LED used for the optical and electrical relaxation oscillator reviewed in the previous section (see Fig. 1.12a). Also, curve A was shifted toward lower voltages compared with curve B. This shift was particularly remarkable when the forward current was 30 mA and higher. The reason for this is that the electron number density in the conduction band is reduced because a population inversion occurs around a forward current of 30 mA, and electrons are consumed for stimulated emission. As a result, the voltage required for injecting the same number of electrons is decreased. On the other hand, when the forward current is increased further, the amount of shift is reduced. This is because the probability of stimulated emission recombination driven by spontaneous emission is increased as the forward current increases, and as a result, the voltage drop due to stimulated emission recombination becomes relatively small.

(3) For evaluating optical amplification characteristics, the relationship between the incident light power P and the current variation ΔI was measured. In a conventional Si-PD, only light absorption is used for photoelectric conversion. In the present Si-PD, however, because stimulated emission is also used, the current variation ΔI depends on the number of electron–hole pairs and varies due to stimulated emission, which is expressed as

$$\Delta I = (eP/h\nu)\,(G-1).\tag{1.4}$$

Here, e is the electron charge, $h\nu$ is the photon energy, and G is the stimulated emission gain. Figure 1.40 shows the relationship between the incident

Fig. 1.40 Relationship between the incident light power at a wavelength of 1.32 μm and the current variation. The *closed circles* and *closed triangles* are for forward current densities of 60 mA/cm2 and 9 A/cm^2, respectively. The *curves A* and *B* show calculated curves fitted to the experimental values

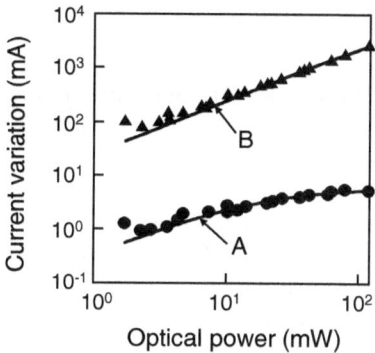

light power at 1.32 μm-wavelength and the current variation ΔI. Similarly to Fig. 1.38, the closed circles and closed triangles show the measurement results for forward current densities of 60 mA/cm^2 and 9 A/cm^2, respectively. For the forward current density of 60 mA/cm^2, ΔI saturates as the incident light power increases; whereas for the forward current density of 9 A/cm^2, ΔI does not saturate but increases linearly. The curves A and B show calculation results fitted to the experimental results using (1.4). The stimulated emission gain G depends on the incident light power as $\exp\left[g/\left(1 + P/P_s\right)\right]$ [96], where g is the small-signal gain coefficient, and P_s is the saturation power. For the fitted results, these values are $g = 3.2 \times 10^{-4}$ and $P_s = 17$ mW in the case of curve A, and $g = 2.2 \times 10^{-2}$ and $P_s = 7.1 \times 10^2$ mW in the case of curve B. The increases in g and Ps as the forward current increases are due to the increasing number of electrons recombining with holes. The experimental values and calculated values show good agreement, confirming that the remarkable increase in photosensitivity shown by the closed circle and closed triangle in Fig. 1.38 is due to optical amplification as a result of stimulated emission.

(4) The photosensitivity was measured when a reverse-bias voltage V_r was applied, while keeping the incident light power fixed. The purpose of this measurement was to verify that the remarkable increases in photosensitivity were due to optical amplification based on stimulated emission under forward current injection. From the measurement results of the relationship between the reverse-bias voltage and the photosensitivity, the maximum photosensitivity was found to be about 3×10^{-4} A/W, which is about 1/300 of the value (0.10 A/W) indicated by the closed triangle in Fig. 1.38. When $V_r = 0$, the photosensitivity was too low to be measured. When V_r was low, for example, $V_r = -1$ V, the photosensitivity was about 2.5×10^{-5} A/W, and therefore, the value of 0.10 A/W above shows that the photosensitivity at $V_r = 0$ was amplified by a factor of at least 4000. Because the photosensitivity saturated as V_r increased, no electron avalanche effect [97] occurred. This means that there is no contribution from an avalanche effect in the remarkable increase in photosensitivity observed when a forward current was applied. Therefore, it was confirmed that the remarkable increase in photosensitivity in the fabricated Si-PD was due to optical amplification based on stimulated emission.

1.11 Summary

This article reviewed light emitting diodes and lasers made of indirect-transition-type Si bulk crystals in which the light emission principle was based on dressed photons. After presenting physical pictures of dressed photons and dressed-photon–phonons, the principle of light emission by using dressed-photon–phonons was reviewed. A novel phenomenon named photon breeding was also reviewed. Next, the fabrication and operation of infrared and visible light emitting diodes and lasers were described, in which the role of coherent phonons in these devices is discussed. Finally, light emitting diodes using SiC and ZnO crystals were described, and other relevant Si devices (a relaxation oscillator and an infrared photodetector with optical amplification) were also reviewed.

References

1. F. Yang, M. Willkinson, E.J. Austin, K.P. OfDonnell, Phys. Rev. Lett. **70**, 323 (1993)
2. Z. Alferov, Semiconductors **32**, 1 (1988)
3. R.A. Milano, P.D. Dapkus, G.E. Stillman, IEEE Trans. Electron. Devices **29**, 266 (1982)
4. U.S. Department of Health and Human Services, Public Health Service, National Inst. Health, National toxicology program (ed.), *NTP technical report on the toxicology and carcinogenesis studies of indium phosphide* (U.S. Department of Health and Human Services, Washington, D.C., 2012) NTP TR 499
5. K.T. Delaney, P. Rinke, C.G. Van de Walle, Appl. Phys. Lett. **94**, 191109 (2009)
6. K.D. Hirschman, L. Tysbekov, S.P. Duttagupta, P.M. Fauchet, Nature **384**, 338 (1996)
7. Z.H. Lu, D.J. Lockwood, J.-M. Baribeau, Nature **378**, 258 (1995)
8. L. Dal Negro, R. Li, J. Warga, S.N. Beasu, Appl. Phys. Lett. **92**, 181105 (2008)
9. T. Komoda, Nucl. Instrum. Methods Phys. Res. Sect. B, Beam Interact. Mater. Atoms **96**, 387 (1995)
10. S. Yerci, R. Li, L. Dal Negro, Appl. Phys. Lett. **97**, 081109 (2010)
11. S.K. Ray, S. Das, R.K. Singha, S. Manna, A. Dhar, Natnoscale Res. Lett. **6**, 224 (2011)
12. M. Ohtsu (ed.), *Progress in Nanophotonics*, vol. 1 (Springer, Berlin, 2011), pp. 1–4
13. M. Ohtsu, *Dressed Photon* (Springer, Berlin, 2013)
14. M.A. Tran, T. Kawazoe, M. Ohtsu, Appl. Phys. A **115**, 105 (2014)
15. R.J. Van Overstraeten, P. Mertens, Sold-State Electron. **30**, 1077 (1987)
16. M.G. Bernald, G. Duraffourg, Phys. Status Solidi **1**, 699 (1961)
17. A. Einstein, P. Ehrenfest, Z. Phys. **19**, 301 (1923)
18. W. Goldammer, W. Ludwig, Z. Zierau, Phys. Rev. B **36**, 4624 (1987)
19. T. Kawazoe, M.A. Mueed, M. Ohtsu, Phys. Rev. B **104**, 747 (2011)
20. E. Shl, *Nonequilibrium Phase Transitions in Semiconductors* (Springer, Berlin, 1987)
21. R.A. Milano, P.D. Dapkus, G.E. Stillman, IEEE Trans. Electron. Devices **29**, 266 (1982)
22. Y. Tanaka, K. Kobayashi, J. Microsc. **229**, 228 (2008)
23. K. Huang, A. Rhys, Proc. R. Soc. Lond. A **204**, 406 (1950)
24. H. Zhao, H. Kalt, Phys. Rev. B **68**, 125309 (2003)
25. T.W. Hagler, K. Pakbaz, K.F. Voss, A.J. Heeger, Phys. Rev. B **44**, 8652 (1991)
26. M. Yamaguchi, T. Kawazoe, M. Ohtsu, Appl. Phys. A **115**, 119 (2014)
27. H. Palevsky, D.J. Hughes, W. Kley, E. Tunkelo, Phys. Rev. Lett. **2**, 258 (1959)
28. J.R. Goldman, J.A. Prybyla, Phys. Rev. Lett. **72**, 1364 (1994)
29. A.J. Sabbah, D.M. Riffe, Phys. Rev. B **66**, 165217 (2002)

30. T. Kawazoe, K. Kobayashi, S. Takubo, M. Ohtsu, J. Chem. Phys. **122**, 024715 (2005)
31. P. Giannozzi, S. Gironcoli, P. Pavone, S. Baroni, Phys. Rev. B **43**, 7231 (1991)
32. S. Nomura, T. Kobayashi, Phys. Rev. B **45**, 1305 (1992)
33. D. Bermejo, M. Cardona, J. Non-Cryst. Solids **32**, 405 (1979)
34. T. Prokofyeva, M. Seon, J. Vanbuskirk, M. Holtz, S.A. Nikishin, N.N. Faleev, H. Temkin, S. Zollner, Phys. Rev. B, Rev. B, **63**, 125313 (2001)
35. D.M. Riffe, A.J. Sabbah, Phys. Rev. B **76**, 085207 (2007)
36. M. Hase, M. Katsuragawa, A.M. Constantinescu, H. Petek, Nat. Photonics Lett. **6**, 243 (2012)
37. A.V. Kuznetsov, C.J. Stanton, Phys. Rev. Lett. **73**, 3243 (1994)
38. N. Wada, M.-A. Tran, T. Kawazoe, M. Ohtsu, Appl. Phys. A **115**, 113 (2014)
39. Y.-X. Yan, E.B. Gamble Jr, K.A. Nelson, J. Chem. Phys. **83**, 5391 (1985)
40. S.I. Kudryashov, M. Kandyla, C.A.D. Roeser, E. Mazur, Phys. Rev. B **75**, 085207 (2007)
41. T. Soma, Phys. Stat. Sol. B **99**, 701 (1980)
42. T.R. Hart, R.L. Aggarwal, B. Lax, Phys. Rev. B **1**, 638 (1970)
43. D. Liang, J.E. Bowers, Nat. Photonics **4**, 511 (2010)
44. H. Rong, R. Jones, A. Liu, O. Cohen, D. Hak, A. Fang, M. Paniccia, Nature **433**, 725 (2005)
45. S. Saito, Y. Suwa, H. Arimoto, N. Sakuma, D. Hisamoto, H. Uchiyama, J. Yamamoto, T. Sakamizu, T. Mine, S. Kimura, T. Sugawara, M. Aoki, Appl. Phys. Lett. **95**, 241101 (2009)
46. T. Kawazoe, M. Ohtsu, K. Akahane, N. Yamamoto, Appl. Phys. B **107**, 659 (2012)
47. W.J. Choi, P.D. Dapkus, J.J. Jewell, IEEE Photonics Tech. Lett. **11**, 1572 (1999)
48. T. Tanbun-Ek, N.A. Olsson, R.A. Logan, K.W. Wecht, A.M. Sergent, IEEE Photonics Tech. Lett. **3**, 103 (1991)
49. Zh.I. Alferov, Semiconductors **32**, 1 (1998)
50. M. Bhatnagar, B.J. Baliga, IEEE Trans. Electron. Devices **40**, 645 (1993)
51. G. Ziegler, P. Lanig, D. Theis, C. Weyrich, IEEE Trans. Electron Devices **30**, 277 (1983)
52. J.A Edmond, H.-S. Kong, C.H. Carter Jr, Phys. B **185**, 453 (1993)
53. T. Kawazoe, M. Ohtsu, Appl. Phys. A **115**, 127 (2014)
54. C. Persson, U. Lindefelt, Phys. Rev. B **54**, 10257 (1996)
55. W. Breinl, J. Friedrich, D. Haarer, Chem. Phys. Lett. **106**, 487 (1984)
56. H. Nienhaus, T.U. Kampen, W. Monch, Surf. Sci. **324**, L328 (1995)
57. Y.-S. Choi, J.-W. Kang, D.-K. Hwang, S.-J. Park, IEEE Trans. Electron. Devices **57**, 26 (2010)
58. D. Seghier, H.P. Gislason, J. Mater. Sci., Mater. Electron. **19**, 687 (2008)
59. A. Tsukazaki, A. Ohtomo, T. Onuma, M. Ohtsni, T. Makino, M. Sumiya, K. Ohtani, S.F. Chichibu, S. Fuku, Y. Segawa, H. Ohno, H. Koinuma, M. Kawasaki, Nat. Mater. **4**, 42 (2005)
60. A. Tsukazaki, M. Kubota, A. Ohtomo, T. Onuma, K. Ohtani, H. Ohno, S.F. Chichibu, M. Kawasaki, Jpn. J. Appl. Phys. **44**, L643 (2005)
61. W.Z. Xu, Z.Z. Ye, Y.J. Zeng, L.P. Zhu, B.H. Zhao, L. Jiang, J.G. Lu, H.P. He, S.B. Zhang, Appl. Phys. Lett. **88**, 173506 (2006)
62. Z.P. Wei, Y.M. Lu, D.Z. Shen, Z.Z. Zhang, B. Yao, B.H. Li, J.Y. Zhang, D.X. Zhao, X.W. Fan, Z.K. Tang, Appl. Phys. Lett. **90**, 042113 (2007)
63. J. Kong, S. Chu, M. Olmedo, L. Li, Z. Yang, J. Liu, Appl. Phys. Lett. **93**, 132113 (2008)
64. A. Nakagawa, T. Abe, S. Chiba, H. Endo, M. Meguro, Y. Kashiwaba, T. Ojima, K. Aota, I. Niikura, Y. Kashiwaba, T. Fujiwara, Phys. Status Solidi C **6**, S119 (2009)
65. F. Sun, C.X. Shan, B.H. Li, Z.Z. Zhang, D.Z. Shen, Z.Y. Zhang, D. Fan, Opt. Lett. **36**, 499 (2011)
66. Z.Q. Chen, T. Sekiguchi, X.L. Yuan, M. Maekawa, A. Kawasuso, J. Phys.: Condens. Matter **16**, S293 (2004)
67. K. Kitamura, T. Kawazoe, M. Ohtsu, Appl. Phys. B **107**, 293 (2012)
68. T. Sekiguchi, S. Miyashita, K. Obara, T. Shishido, N. Sakagami, J. Cryst. Growth, **214/215**, 72 (2000)
69. Ü.Ö zgur, Y.I. Alivov, C. Liu, A. Teke, M.A. Reshchikov, S. Doğan, V. Avrutin, S.-J. Cho, H. Morkoc, J. Appl. Phys. **98**, 041301 (2005)
70. M.S. Kim, K.G. Yim, J.Y. Leem, D.Y. Lee, J.S. Kim, J.S. Kim, J. Phys. Soc. **58**, 821 (2011)
71. H. Mach, C.R. Middaugh, Anal. Biochem. **222**, 323 (1994)

72. O. Svelto, *Principles of Lasers*, 2nd edn. (Plenum Press, New York, 1982)
73. N. Wada, T. Kawazoe, M. Ohtsu, Appl. Phys. B **108**, 25 (2012)
74. M. Meier, S. Karg, W. Riess, J. Appl. Phys. **82**, 1961 (1997)
75. J. Katz, S. Margalit, C. Harder, D. Wilt, A. Yariv, J. Quantum Electron. **17**, 4 (1981)
76. G. Bjork, Yamamoto. J. Quantum Electron. **27**, 2386 (1991)
77. E. Finkman, M.D. Sturge, R. Bhat, J. Lumin. **35**, 235 (1986)
78. V. Zwiller, T. Aichele, W. Seifert, J. Persson, O. Benson, Appl. Phys. Lett. **82**, 1509 (2003)
79. K. Okamoto, I. Niki, A. Scherer, Y. Narukawa, T. Mukai, Y. Kawakami, Appl. Phys. Lett. **87**, 071102 (2005)
80. M.E. Levinshtein, S.L. Rumyantsev, M. Shur, *Handbook Series on Semiconductor Parameters*, vol. 1 (World Scientific Publishing, Singapore, 1996)
81. A. Loudon, P.A. Hiskett, G.S. Buller, Opt. Lett. **27**, 219 (2002)
82. C. Cremer, N. Emeis, M. Schier, G. Heise, G. Ebbinghaus, L. Stoll, IEEE Photonics Tech. Lett. **4**, 108 (1992)
83. A.F. Phillips, S.J. Sweeney, A.R. Adams, P.J.A. Thijs, IEEE J. Sel. Topics Quantum Electron. **5**, 401 (1999)
84. J.E. Carey, C.H. Crouch, M. Shen, E. Mazur, Opt. Lett. **30**, 1773 (2005)
85. M.W. Geis, S.J. Spector, M.E. Grein, R.T. Schulein, J.U. Yoon, D.M. Lennon, C.M. Wynn, S.T. Palmacci, F. Gan, F.X. Kartner, T.M. Lyszczarz, Opt. Express **15**, 16886 (2007)
86. M.W. Geis, S.J. Spector, M.E. Grein, R.T. Schulein, J.U. Yoon, D.M. Lennon, F. Gan, F.X. Kartner, T.M. LyszczarzI, IEEE Photonics Technol. Lett. **19**, 152 (2007)
87. T. Baehr-Jones, M. Hochberg, A. Scherer, Opt. Express **16**, 1659 (2008)
88. H. Chen, X. Luo, A.W. Poon, Appl. Phys. Lett. **95**, 171111 (2009)
89. M. Lee, C. Chu, Y. Wang, Opt. Lett. **26**, 160 (2001)
90. M. Cassalino, L. Sirleto, L. Moretti, M. Gioffre, G. Coppola, Appl. Phys. Lett. **92**, 251104 (2008)
91. T. Tanabe, K. Nishiguchi, E. Kuramochi, M. Notomi, Appl. Phys. Lett. **96**, 101103 (2010)
92. B. Shi, X. Liu, Z. Chen, G. Jia, K. Cao, Y. Zhang, S. Wang, C. Ren. J. Zhao, Appl. Phys. B **93**, 873 (2008)
93. H. Tanaka, T. Kawazoe, M. Ohtsu, Appl. Phys. B **108**, 51 (2012)
94. S. Yukutake, T. Kawazoe, T. Yatsui, W. Nomura, K. Kitamura, M. Ohtsu, Appl. Phys. B **99**, 415 (2010)
95. J.D. Schaub, J. Lightwave Technol. **19**, 272 (2001)
96. T. Saitoh, T. Mukai, IEEE J. Quantum Electron. **23**, 1010 (1987)
97. R.J. McIntyre, IEEE Trans. Electron. Devices **13**, 164 (1966)

Chapter 2
Theoretical Analysis on Optoelectronic Properties of Organic Materials: Solar Cells and Light-Emitting Transistors

Hiroyuki Tamura

Abstract This chapter reviews our recent theoretical studies on optoelectronic properties of molecular condensates for organic solar cells and light-emitting devices. The following three sections comprises this chapter: (1) photo-induced charge separation in organic solar cells, (2) exciton diffusion length and charge mobility in interpenetrating organic solar cells, and (3) organic light-emitting transistors. We discuss the physical origins underlying organic photovoltaics and light-emissions for rationalizing the experimental measurements.

2.1 Introduction

This chapter introduces our recent theoretical studies on optoelectronic properties of molecular condensates, namely organic solar cells and organic light-emitting transistors. The electronic structures of molecular condensates can properly be described by site-based model Hamiltonians which are parametrized by first principles calculations. This approach is well suited for analyzing the physical picture of charge transports and exciton dynamics in molecular condensates.

2.2 Photo-induced Charge Separation in Organic Solar Cells

The potential advantages of organic solar cells are thought to be their low production cost which potentially shortens the energy payback time, as well as the properties peculiar to organic materials such as light weight and flexibility which are favorable for installing on walls, curved surfaces, and for various niche applications including mobile and wearable devices [1]. At present, the energy conversion efficiency of organic solar cells is at most ~10 %, and thus less efficient than the conventional

H. Tamura (✉)
WPI-Advanced Institute for Material Research (AIMR), Tohoku University,
2-1-1 Katahira Aobaku, Sendai 980-8577, Japan
e-mail: hiroyuki@wpi-aimr.tohoku.ac.jp

© Springer International Publishing Switzerland 2015
M. Ohtsu and T. Yatsui (eds.), *Progress in Nanophotonics 3*,
Nano-Optics and Nanophotonics, DOI 10.1007/978-3-319-11602-0_2

silicon solar cells. Nevertheless, growing energy conversion efficiency of organic solar cells is expected to enhance their practical usefulness.

Organic solar cells convert sunlight into electricity through photoabsorption, exciton diffusion to the donor-acceptor interface, exciton dissociation at the interface, charge transports to the electrodes, and charge extraction [1–33]. The donor and acceptor materials of standard bulk heterojunction organic solar cells are typically π-conjugated polymers, e.g., poly-3-hexylthiophene (P3HT), and fullerene derivatives, e.g., [6,6]-phenyl-C_{61} butyric acid methyl ester (PCBM). Small π-conjugated molecules such as porpyrin derivatives and phthalocyanine are also employed for donor materials.

The optoelectronic properties of the donor and acceptor materials, such as absorption spectra, donor-acceptor band offset, and charge mobility can be controlled by chemical modifications based on appropriate design rules. For example, low band gap donor molecules have been developed for efficiently utilizing long wavelength region of sunlight spectrum [4, 16]. Besides, the LUMO level of acceptor molecules have been controlled by chemical modification so as to increase the open circuit voltage (Fig. 2.1). Electronic structure calculations can contribute to the prediction of such properties. Moreover, the theoretical analysis on the dynamics of exciton and charge carriers can provide useful insight into the mechanisms of photovoltaics.

In the following sections, we introduce our recent theoretical studies on organic photovoltaics.

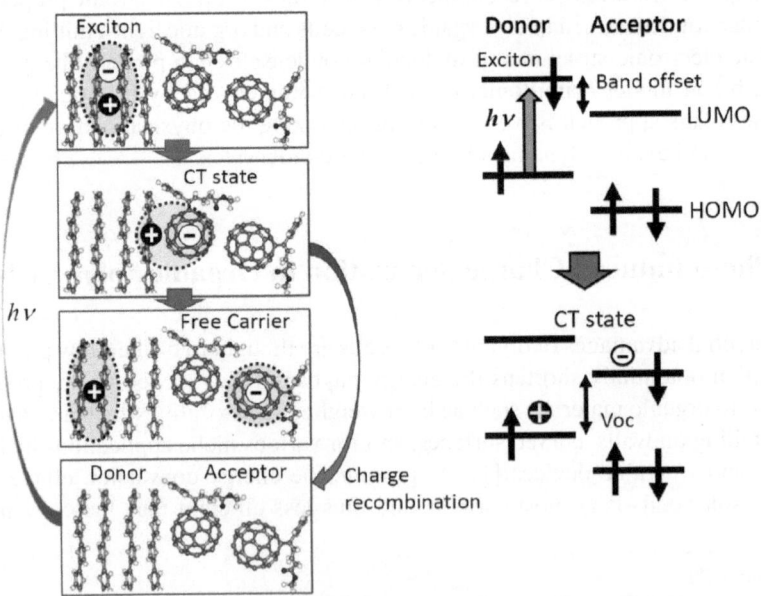

Fig. 2.1 Diagram of photo-induced charge separation at donor-acceptor interfaces

2.2.1 Charge Separation at Donor-Acceptor Interfaces

Charge separation at donor-acceptor heterojunctions is a key process that determines the energy conversion efficiency of organic solar cells [1–13, 15–18]. The
photo-generated exciton is thought to primarily decay to a bound electron-hole pair
localized at the donor-acceptor interface referred to as charge transfer (CT) state
[1–3] (Fig. 2.1). For the photo-current conversion, the electron-hole pair should separate into free carriers overcoming the Coulomb attraction. The charge separation can
compete with the electron-hole charge recombination, i.e., decay from the interfacial
CT state to the ground state [4]. Thus, as the charge separation becomes faster, the
charge recombination decreases and the internal quantum efficiency can be improved.

The difference in the chemical potential between the anode and cathode induces an
internal electric field of typically ~ 10 V/μm in the organic layer, thereby providing a
driving force for the charge separation [17] (Fig. 2.2). The internal quantum efficiency
can be improved by increasing the internal electric field [17].

Since the dielectric constant of organic materials is generally small ($\epsilon_r = 3 \sim 4$),
the strong electron-hole Coulomb attraction stabilizes the interfacial CT state [1–4].
The potential barrier for the dissociation of point charges is typically ~ 0.4 eV, which
is much higher than the thermal energy at room temperature (~ 0.026 eV).

How does the electron-hole pair separate into free carriers overcoming the
Coulomb attraction? To answer this questions, the following effects have been
pinpointed in recent investigations: First, the delocalization of electron and hole
[10, 13, 21, 27, 29] and second, the excess energy of the photo-generated exciton, entailing the so-called hot exciton dissociation mechanism [12, 15, 16, 18, 27].
We theoretically clarify how the interfacial CT state separates into free carriers
at polymer/fullerene donor/acceptor interfaces (Fig. 2.3). A detailed microscopic

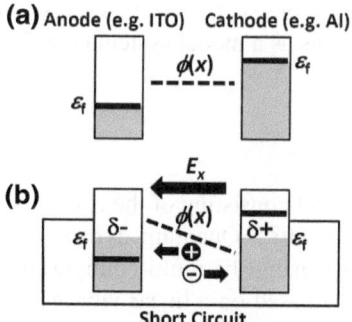

Fig. 2.2 Diagram of the internal electric field, E_x, of organic solar cells induced by the chemical
potential difference of the anode and cathode; **a** isolated electrodes, i.e., open circuit and **b** charge
transferred electrodes via the short circuit. The *gray* area indicates the occupied energy levels,
where the *bold solid lines* indicate the Fermi level of isolated electrodes. The *dashed lines* depict
the Coulomb potential, $\phi(x)$, induced by the electrodes [26]

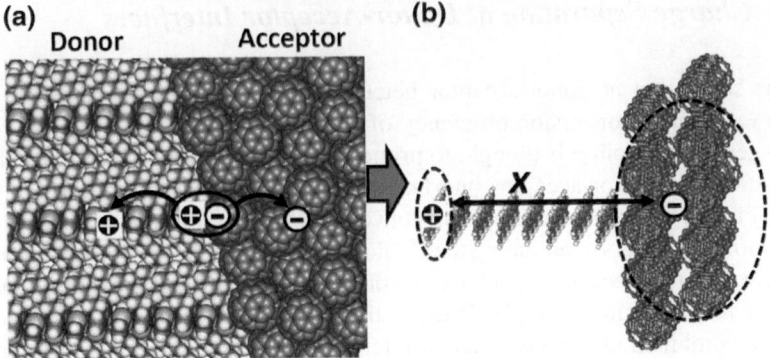

Fig. 2.3 **a** Schematic illustration of a crystalline domain of the P3HT/Fullerene donor/acceptor interface, considering a lamellar stacking of P3HT and a hexagonal close-packed cluster of fullerene. **b** Model systems consisting of π-stacked oligothiophenes (T_n: $n = 5 \sim 13$) and a C_{60} cluster. Here, the electron-hole distance, x, is defined as the center-to-center distance between a T_n molecule and the C_{60} molecule at the center of the first interface layer

analysis is carried out using quantum dynamical simulations with a parametrisation based on density functional theory (DFT) and time-dependent DFT (TDDFT) calculations.

Experimentally, charge separation in P3HT/PCBM [9] and related blends [16] has been observed on a time scale of < 100 fs by pump-probe studies. The observed sub-picosecond charge separations imply a coherent nature of the delocalized charges and excitons. In such systems, kinetic models such as Marcus theory [34] and the Onsager-Braun model [35] do not necessarily hold, such that non-perturbative quantum dynamical analysis is necessary. We explicitly account for electron-phonon (vibronic) couplings, which generally play an essential role in the charge and exciton transfers in organic semiconductors. To provide realistic parameters representing polymer/fullerene heterojunctions such as P3HT/PCBM, we consider oligothiophene (T_n)/fullerene (C_{60}) interfaces as a model system (Fig. 2.3).

2.2.2 Methods

We employ a site-based model consisting of the exciton (XT), charge transfer (CT), and charge separated (CS) states. Quantum dynamics calculations of the charge separation were carried out using the multi-configuration time-dependent Hartree (MCTDH) method [36, 37] based on a linear vibronic coupling (LVC) model in a site-based diabatic representation. The Hamiltonian takes the following form:

$$
\begin{aligned}
H = {} & h_{XT}(\mathbf{x})|XT\rangle\langle XT| + h_{CT}(\mathbf{x})|CT\rangle\langle CT| \\
& + \sum_n h_{CS_n}(\mathbf{x})|CS_n\rangle\langle CS_n| + \gamma(|XT\rangle\langle CT| + h.c.) \\
& + t(|CT\rangle\langle CS_1| + h.c.) + \sum_{nn'} t|CS_n\rangle\langle CS_{n'}|
\end{aligned}
\tag{2.1}
$$

$$h_\xi(x) = \sum_i \omega_i/2(x_i^2 + p_i^2) + \sum_i \kappa_i^{(\xi)} x_i + \epsilon^{(\xi)} \tag{2.2}$$

where $\xi = (\text{XT, CT, CS}_n)$ denotes the exciton, the CT state, and the CS states, respectively; $x = x_i$ collectively denotes a set of intramolecular phonon modes; $h_\xi(x)$ are the Hamiltonians of these harmonic modes in the respective states, and ω_i and p_i are the frequency and momentum of the vibrational modes, respectively. Further, $\epsilon^{(\xi)}$ are the respective on-site potentials. The delocalized initial exciton involves an additional coupling term, $J(|XT_1\rangle\langle XT_2| + h.c.)$. The XT-CT coupling γ, the CT integral t, the exciton coupling J, and the spectral density of vibronic couplings κ_i, were determined by DFT and linear response time-dependent DFT (TDDFT) calculations, in conjunction with the quasi-diabatization scheme as follows [24, 26, 27]:

1. We prepare the set of reference wavefunctions, Φ_{refI}, that possess pure exciton and CT characters.
2. We calculate the adiabatic states of the donor-acceptor interface at a given interface structure, for which the intermolecular interaction can induce mixing of exciton and CT characters.
3. The diabatic wavefunctions are represented by a linear combination of adiabatic wavefunctions, Ψ_i, i.e., $\Phi_I = \sum_i C_{Ii}\Psi_i$, $C_{Ii} = \langle\Psi_i|\Phi_{refI}\rangle$

That is, the adiabatic states from the TDDFT calculations are considered as basis functions for expanding the diabatic states. In this study, the reference wavefunctions were calculated at a sufficiently long intermolecular distance at which the adiabatic states correspond to the pure exciton and CT states. Then, for evaluating the overlap between Ψ_i and Φ_{refI}, the intermolecular distance is changed while the eigen vectors of the reference wavefunctions are fixed. Although TDDFT does not provide wavefunctions in a rigorous sense, quasi-wavefunctions are constructed based on a superposition of the single excitations, where electron exchanges are taken into account in the same way as the Slater determinants. Here, the minor components, namely de-excitation components in TDDFT, are neglected in the quasi-diabatization scheme. The diabatic potentials and couplings are evaluated from the diabatic wavefunctions as follows:

$$\begin{aligned} E_i &= \langle\Phi_i|h|\Phi_i\rangle \\ E_j &= \langle\Phi_j|h|\Phi_j\rangle \\ V &= \langle\Phi_i|h|\Phi_j\rangle \end{aligned} \tag{2.3}$$

where h is the electronic Hamiltonian.

DFT tends to underestimate the energy of CT states, and thus the excited states were calculated using the long-range corrected TDDFT (LC-TDDFT) [38], which can fairly well describe CT states. The XT–CT offset is defined as $\Delta E_{XT-CT} = \epsilon_{XT} - \epsilon_{CT}$. In this study, we consider a reasonable range of ΔE_{XT-CT} based on the LC-TDDFT calculations for T_n/C_{60} complexes of different π-conjugation lengths. The vibrational modes were reduced to a limited number of effective modes, which reproduce the short-time dynamics and the reorganization energy of the whole system [22].

2.2.3 Charge Transfer Integral and Exciton Coupling for π-Stacked Oligothiophene Molecules

We carried out DFT and TDDFT calculations to estimate a reasonable range of the transfer integral and the exciton coupling for π-stacked oligo-thiophene (T_n) molecules. According to the DFT calculations by Dag et al. for the lamellar structure of P3HT [39], the shifted configuration is more stable than the aligned one owing to the steric effect of the hexyl groups. We considered aligned, shifted, and half-shifted π-stacking configurations of a T_n dimer. Figure 2.4 shows the transfer integral and the exciton coupling as a function of the center-to-center distance, R. The transfer integral at R of $3.5 \sim 3.8$ Å is found to be $0.2 \sim 0.07$ eV depending on the stacking configurations and π-conjugation length.

2.2.4 Potential Barrier for Electron-Hole Separation

The potential between a hole distributed over the oligothiophene moiety and an electron delocalized over the hexagonal close-packed C_{60} clusters is calculated based on the tight-binding model as a function of the electron-hole distance, x (Fig. 2.5) [27].

The tight-binding Hamiltonian describing the electron delocalized over the C_{60} cluster reads:

$$H(x) = \sum_{i=1}^{N} \epsilon_i(x)|i\rangle\langle i| + \sum_{i=1,j=1}^{N} t_{ij}|i\rangle\langle j| \qquad (2.4)$$

where $\epsilon_i(x)$ is the on-site potential at each C_{60} molecule, t_{ij} is the charge transfer integral between the C_{60} molecules, and N is the number of C_{60} molecules. Here, only t_{ij} of nearest neighbors are considered. t_{ij} is assumed to be 0.05 eV based on the DFT calculations [40]. The on-site potential, ϵ_i, is determined considering the Coulomb interaction between a point electron at a C_{60} site and a hole described as a line-shaped charge distribution on a thiophene chain [26]. The hole distribution corresponds to the effective π-conjugation length of a polymer chain. The DFT calculations indicated that the potential barrier to charge separation decreases as the π-conjugation length of the donor molecule increases, owing to the delocalization

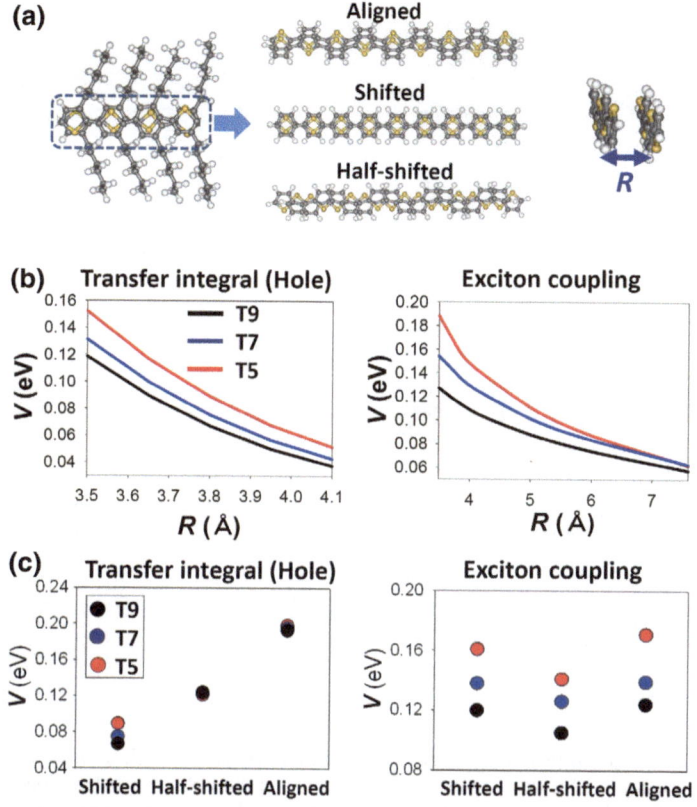

Fig. 2.4 a π-stacking configurations of T_n dimer. **b** Transfer integral and exciton coupling as a function of the intermolecular distance, R, where the shifted configuration is considered. **c** Transfer integral and exciton coupling for the shifted, half-shifted, and aligned configurations at R of 3.8 Å

of the hole [26]. The effective π-conjugation length of polymers is affected by the crystallinity at donor/acceptor interfaces, where annealing is known to enhance the crystallization [9, 19].

The electric field from the electrodes, E_x, is assumed to be perpendicular to the π-conjugation plane of donor. We consider E_x of 10 V/µm and the dielectric constant, ϵ_r, of 4 (representative of polythiophene). The lowest eigen state of the tight-binding Hamiltonian (2.1) as a function of the electron-hole distance, x, corresponds to the potential curve for the charge separation. Here we assume the initial electron-hole distance of 7 Å and the T_n π-stacking distance of 3.8 Å.

We calculate the electron-hole potential considering various sizes of fullerene clusters, where small versus large clusters represent disordered versus crystalline domains, respectively. Our calculations indicate that the potential barrier becomes shallower as the size of the fullerene cluster becomes larger (Fig. 2.6), since the electron can delocalize over an increasing number of fullerene sites [27].

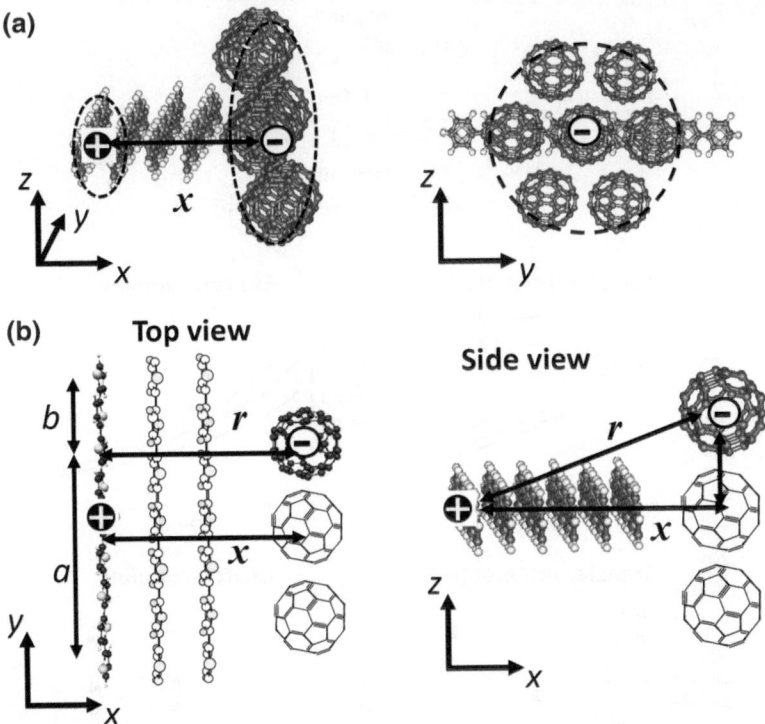

Fig. 2.5 a Schematic illustration of the T_n/C_{60} donor-acceptor model. x denotes the distance between the electron delocalized over the C_{60} cluster and the hole distributed along a T_n chain. **b** Definitions of the distance, r_i, between the ith C_{60} site and a T_n molecule, where the *accompanying arrow* indicates the perpendicular line from the C_{60} to T_n. Here, $a + b$ is the effective π-conjugation length of T_n (2.1)

To simulate the charge separation dynamics at a molecular level, the potential curve is mapped onto the on-site energies of the CS states consisting of an electron on the fullerene cluster and a hole on a polymer chain. Figure 2.7b, c show snapshots of the population distribution over the respective states during a 400 fs interval of the quantum dynamics simulations [27]. The CT population rises within a few tens of femtoseconds (fs), and in turn the CS states are populated immediately (Fig. 2.7b–e). Such ultrafast charge separations were indeed observed in the pump-probe experiments of polymer/fullerene photovoltaic systems [9, 16]. As the inter-donor transfer integral increases, the charge separation becomes more efficient, where the hole can be delocalized over many molecules and can penetrate into the high potential sites, i.e., through-barrier tunneling can play a role. The CS states beyond the potential barrier (typically CS_n with n > 10–15, see Fig. 2.7a) are populated within $100 \sim 200$ fs, which is much faster than the typical time scale of charge re-combination (i.e., a few hundred picoseconds) [4]. Hereafter, the sum of the CS populations beyond the barrier, i.e., the free carrier yield, is denoted as η_{FC}. The interfacial CT population

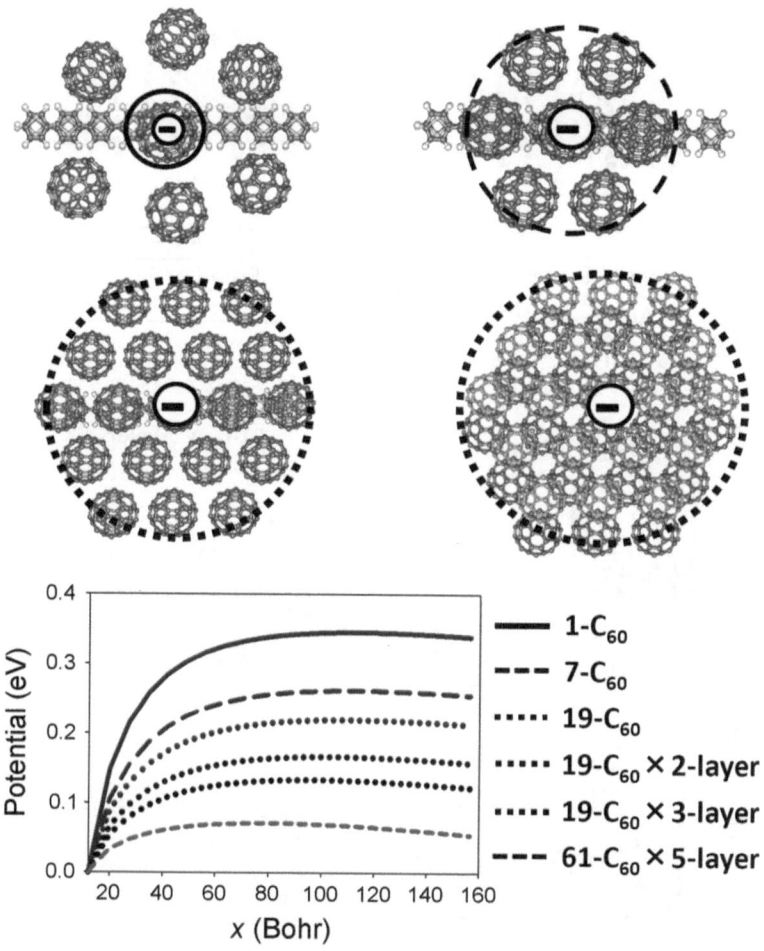

Fig. 2.6 Potential curves for the charge separation considering a hole on T_{13} and an electron on various sizes of C_{60} clusters, based on the tight-binding model parametrized by DFT calculations, with a C_{60}–C_{60} charge transfer integral of 0.05 eV, an electric field along the x direction of 10 V/μm, and a dielectric constant of 4

exhibits a plateau after a few hundred fs, corresponding to carriers which remain trapped at the donor-acceptor interface and will eventually decay to the ground state on a longer time scale. That is, the major portion of η_{FC} would be determined by the dynamics during the first few hundred fs.

Our quantum dynamics calculations clearly indicate that the charge separation efficiency increases as the electron is more delocalized within the fullerene condensate, and as the π-conjugation length of the donor, i.e., the hole distribution length, increases. This is because the potential barrier is decreased by the charge delocalization. This trend is generally consistent with the experimental observations that

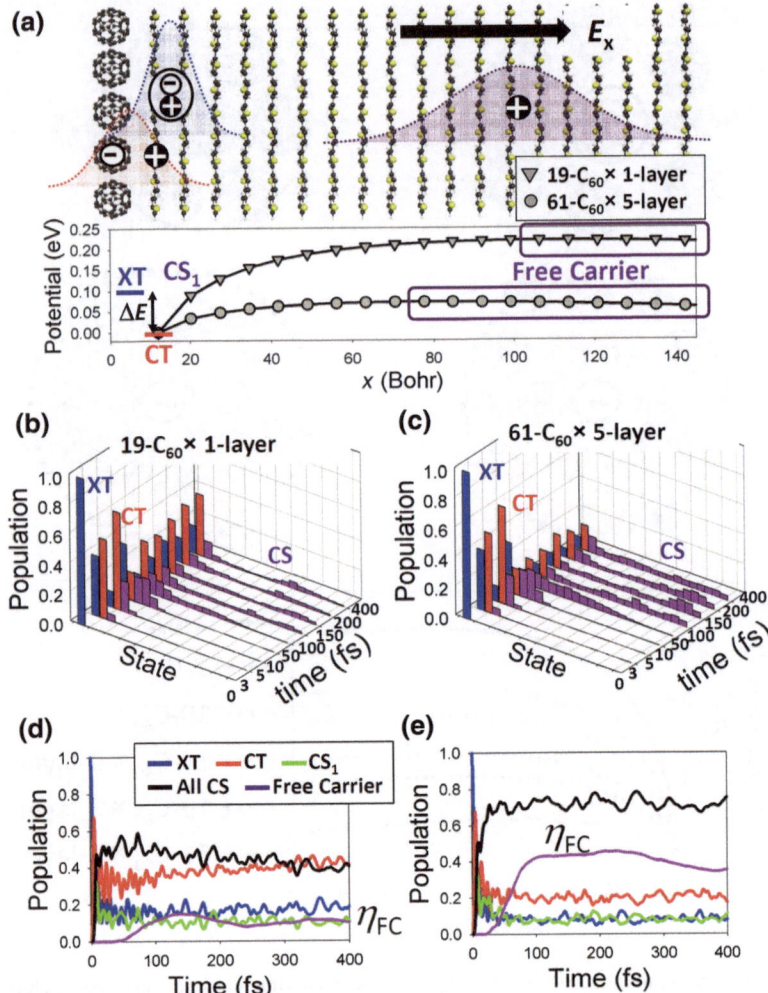

Fig. 2.7 a On-site energies of the respective charge-separated (CS) states, together with schematic illustration of the photo-generated exciton (XT), interfacial CT state, and delocalized polaron on the T_n/C_{60} donor-acceptor model, where the $T_n - T_n$ distance is assumed to be 3.8 Å and an electric field, E_x, of 10 V/μm is applied perpendicular to the π-conjugation plane of the donor. Snapshots of the population distribution on the respective states during the quantum dynamics calculations for **b** disordered and **c** ordered donor/acceptor models. Population plots for these calculations; **d** disordered and **e** ordered models. The sum of CS populations beyond the potential barrier is defined as the free carrier population (see panel (**a**)). The XT–CT coupling of 0.2 eV and the T_n–T_n charge transfer integral of 0.12 eV are considered for all the calculations

indicate improvement of the internal quantum efficiency by increasing the regio-regularity of the donor polymer and by the annealing of polymer/fullerene blends [9] which enhances the crystallization. The experimentally reported barrierless charge

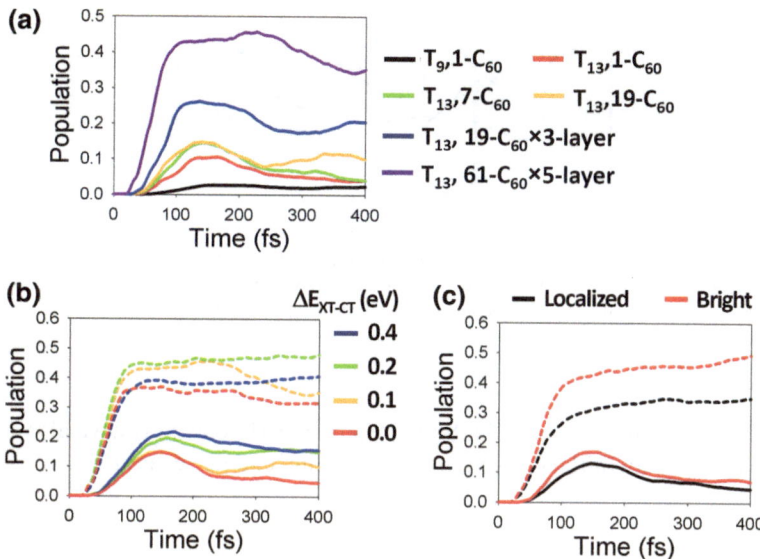

Fig. 2.8 **a** Free carrier yield η_{FC} from the quantum dynamics calculations considering various sizes of C_{60} clusters and π-conjugation lengths of T_n, where ΔE_{XT-CT} is set to 0.1 eV. **b** η_{FC} for various values of ΔE_{XT-CT}. **c** η_{FC} generated from the bright exciton delocalized over two donor molecules, where ΔE_{XT-CT} for a single exciton is set to 0.1 eV and the exciton coupling is 0.15 eV. The *solid* and *dashed lines* indicate η_{FC} using the potentials for the disordered and ordered (delocalized electron) models, respectively. The XT–CT coupling of 0.2 eV and the $T_n - T_n$ charge transfer integral of 0.12 eV are considered for all calculations

separation [7, 8] would be explained by the formation of such favorable domains in bulk heterojunctions.

In order to elucidate the role of hot CT states in the charge separation, we performed quantum dynamics simulations considering various values of the exciton-CT energy offset (ΔE_{XT-CT}). Overall, the calculated η_{FC} increases with increasing ΔE_{XT-CT} [27] (Fig. 2.8). The excess energy of the exciton-CT transition induces vibrational excitations, such that the vibronically excited CT state can become resonant with CS states at energetically high potential sites (Fig. 2.9). On the timescale of a few hundred fs, the CT state is not yet equilibrated, and thus the vibronically hot CT states keep promoting the charge separation.

Besides ΔE_{XT-CT}, the delocalization of the photo-generated exciton on the π-stacked H-aggregate contributes to modifying the excess energy. Bright excitons delocalized over H-aggregates generally possess higher excitation energies and stronger oscillator strengths than localized excitons on a single molecule. For illustration, we consider a coherent bright exciton delocalized over two sites. The coherent bright exciton tends to result in a more efficient charge separation than the localized exciton (Fig. 2.8c) [27]. However, all the excess excitation energy is not necessarily exploited for the charge separation, because the bright exciton can rapidly decay to

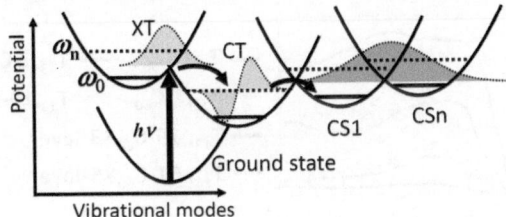

Fig. 2.9 Concept of charge separation mediated by vibronically hot CT states. Potential crossings of XT, CT, and CS states, together with the vibrational wavefunctions on the respective states. The *solid* and *dashed black lines* illustrate the vibrational ground (ω_0) and excited (ω_n) states, respectively

the lower-lying dark exciton [27]. Some recent experiments [14] have also implied the role of coherent initial exciton in enhancing the charge separation.

2.2.5 Summary

The quantum dynamics calculations parametrized by DFT and TDDFT have revealed two main factors that can significantly enhance the free carrier generation. First, the lowering of the Coulomb barrier due to charge delocalization is an indispensable condition for the efficient charge separation. The charge delocalization can be realized in actual organic solar cells by enhancing the crystallization. The barrierless charge separation observed in some experiments [7, 8] can be rationalized by a small potential barrier due to charge delocalization. Second, vibronically hot CT states can enhance the ultrafast charge separation, where the free carrier formation is enhanced substantially as the excess energy increases. The excess energy varies with the exciton-CT offset (ΔE_{XT-CT}) of the donor-acceptor heterojunctions as well as the delocalization of exciton. We found that the vibronically hot CT dissociation is particularly effective when ΔE_{XT-CT} is comparable to the Coulomb barrier. Too much excess energy is not effective for further improvement of free carrier formation [27], such that the use of large-ΔE_{XT-CT} materials is not necessarily a strategy for optimization.

Organic solar cells simultaneously demand large open circuit voltage and long-wavelength absorption for improving the energy conversion efficiency. In this context, efficient charge separation with a small excess energy, i.e., using low-ΔE_{XT-CT} materials, is favorable. While these requirements are incompatible with the presence of a high Coulomb barrier, the barrier can be reduced substantially by charge delocalization. As a result, free carriers can be generated on an ultrafast time scale, without involving higher electronic excitations of the donor species. The present picture is generally applicable for the ultrafast dynamics of photo-induced charge separations in a broad range of heterojunction systems.

2.3 Interpenetrating Organic Solar Cells: Exciton Diffusion Length and Charge Mobility

In organic solar cells, the photo-generated exciton in the condensate of donor molecules should diffuse to the donor-acceptor interface at which the exciton separates into electron and hole. Thus, the exciton diffusion length is one of important factors determining the internal quantum efficiency [9]. Current standard organic solar cells generally consist of bulk heterojunction structures of semiconducting polymers and fullerene derivatives [1], which are favorable for increasing the donor-acceptor interface area. However, bulk heterojunction structures are not necessarily advantageous for controlling the pathway of charge transports from the donor-acceptor interface to the electrodes. The donor-acceptor heterojunction should be designed so as to simultaneously facilitate the charge separation at the interface and the charge transport to the electrodes [41–44]. To achieve this requirement, Matsuo et al. [41, 42] has realized an interpenetrating structure of donor-acceptor heterojunction (Fig. 2.10) consisting of columnar tetrabenzoporphyrin (BP) crystals and bis (dimethylphenylsilylmethyl) [60]fullerene (SIMEF) [45, 46] filling the intercolumnar gaps. Here, the exciton is photo-generated in a column of BP condensate and the charge separation occurs at the BP/SIMEF interface. Then, the hole and electron are transported to the electrodes through the interdigitated BP and SIMEF condensates. The organic solar cells consisting of small donor molecules are also favorable as model systems for understanding the microscopic mechanisms of photovoltaic processes.

We have theoretically analyzed the exciton diffusion length and the charge mobility in the BP and SIMEF crystals [47] by means of density functional theory (DFT) and Fermi's golden rule. Although molecular condensates in the actual organic solar cells would be polycrystalline, we consider the single crystals of BP and SIMEF for fundamental characterizations. Since experimental measurements

Fig. 2.10 Schematic illustration of the interpenetrating heterojunction organic solar cell consisting of BP and SIMEF as electron donor and acceptor, respectively

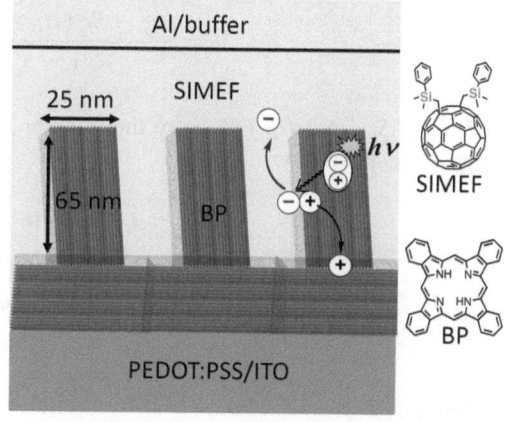

of the diffusion length of singlet excitons are more difficult than that of triplet excitons [48–50], the theoretical prediction of the intrinsic exciton diffusion length is particularly important.

2.3.1 Methods

The charge and exciton transports in organic materials are often described based on the hopping between localized sites. The hopping transport is a thermally activated process, and thus the mobility increases as the temperature becomes higher. Some organic crystals exhibit band-like charge transports [51–55] for which the mobility decreases as the temperature becomes higher. Since the electron-phonon coupling of organic materials is strong, ideal band picture does not hold and the charge transports accompany lattice distortions, so called polaron. The excitons in some molecular assemblies, e.g., in light harvesting complex of photosynthesis [56, 57], are thought to be delocalized over several molecules keeping coherent nature.

The exciton diffusion length and the charge mobility of molecular assemblies can be estimated theoretically based on Fermi's golden rule for hopping transports [34, 58, 59], or by using semi-classical wavepacket dynamics for describing coherent excitons and delocalized polarons [60, 61]. In this section, we focus on the hopping transports. The intermolecular charge hopping rate by the Marcus theory reads [34]:

$$W = \frac{2\pi}{\hbar} |V|^2 \frac{1}{\sqrt{4\pi \lambda k_B T}} \exp\left\{ -\frac{(\Delta E - \lambda)^2}{4\pi \lambda k_B T} \right\} \tag{2.5}$$

where $\hbar$ is the Plank constant, k_B is the Boltzmann constant, T is the temperature, and λ is the reorganization energy. In this study, the driving force, ΔE, is assumed to be zero. The intermolecular electronic coupling, V, may be those for charge transfers as well as for exciton transfers. For charge transfers V reads [59]:

$$V = \frac{J_{ij} - S_{ij}(H_{ii} + H_{jj})/2}{1 - S_{ij}^2} \tag{2.6}$$

where J_{ij}, S_{ij}, and H_{ii} (H_{jj}) are the transfer integral, overlap integral, and on-site energies, respectively.

$$\begin{aligned}
J_{ij} &= \langle \phi_i | h | \phi_j \rangle \\
S_{ij} &= \langle \phi_i | \phi_j \rangle \\
H_{ii} &= \langle \phi_i | h | \phi_i \rangle \\
H_{jj} &= \langle \phi_j | h | \phi_j \rangle
\end{aligned} \tag{2.7}$$

Here, ϕ_i is the HOMO or LUMO orbital of the single molecule relevant to the hole or electron transfer. h is the electronic Hamiltonian of a molecular dimer. These integrals are here evaluated by DFT with the B3LYP functional. This treatment assumes that the on-site energies of ith and jth molecules are identical, where the energy splitting of the bonding and anti-bonding states of a dimer corresponds to $2V$. For systems where the on-site energies of the donor and acceptor are different, the coupling V can be estimated by using the quasi-diabatization scheme as described below [47].

1. Excited states of each single molecule are considered as the reference wavefunctions, Φ_{refI}.
2. Excited states of a dimer of molecules, i.e., adiabatic states Ψ_i, are calculated using linear response time-dependent DFT (TDDFT).
3. The diabatic wavefunctions are represented by a linear combination of the adiabatic wavefunctions, Ψ_i:

The obtained exciton coupling includes the contributions from the Coulomb couplings (Förster transfer) and the electron exchanges (Dexter transfer).

We estimate the angle-dependent charge mobility, μ, as follows [59]:

$$D(\theta_0) = \frac{1}{2} \sum_i W_i P_i R_i^2 \cos^2(\theta_i - \theta_0) \tag{2.8}$$

$$\mu(\theta_0) = \frac{e}{k_B T} D(\theta_0) \tag{2.9}$$

where e is the elementary charge, W_i is the charge transfer rate for the ith intermolecular hopping path, and R_i is the intermolecular center-to-center distance. P_i is the weight of the ith hopping path defined as follows:

$$P_i = \frac{W_i}{\sum_i W_i} \tag{2.10}$$

θ_0 is the angle between the charge transport direction and the a-axis (Fig. 2.11). θ_i is the angle between the ith hopping path and the a-axis.

The exciton transfer rates between BP molecules are also estimated based on the Fermi's golden rule in the same way as 2.5. The reorganization energy is evaluated by geometry optimization of the excited state. The exciton lifetime, τ, is estimated based on the Einstein A-coefficient of spontaneous emission:

$$A = \frac{\omega^3}{3\pi \epsilon_0 \hbar c^3} d^2 = \frac{1}{\tau} \tag{2.11}$$

where ω is the excitation frequency, ϵ_0 is the dielectric constant of vacuum, c is the speed of light, and d is the transition dipole moment. We analyze the angle-dependent exciton diffusion length based on 2.8 and the exciton lifetime, τ:

$$L_D = \sqrt{D\tau} \tag{2.12}$$

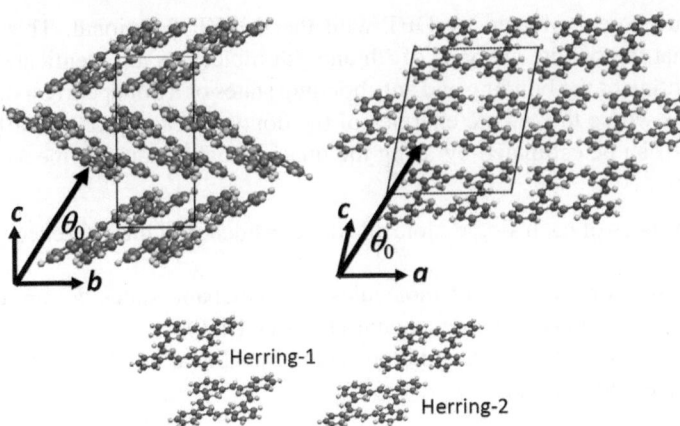

Fig. 2.11 Crystal structure of BP. The *box* indicates the unit cell. The cell parameters are as follows: $a = 12.405$, $b = 6.591$, $c = 14.927$ (Å), $\alpha = 90.0$, $\beta = 101.445$, $\gamma = 90.0$ (degree). Two kinds of herringbone configurations in the crystal are shown

2.3.2 Exciton Diffusion Length and Charge Mobility

The BP crystal takes a herringbone packing structure (Fig. 2.11), where the BP molecules alternately slide along the α-direction, i.e. a staggered herringbone packing. Table 2.1 summarizes the calculated intermolecular electronic couplings of the hole transfers, V_{hole}. The V_{hole} value for the parallel π-stacking H-aggregate (i.e., along b-direction) was especially large, while V_{hole} for the staggered herringbone pairs and the J-aggregate pairs (along a-direction) were generally small. As a result, the hole mobility exhibits a strong anisotropy depending on the transport direction (Fig. 2.3a, b).

The optically bright excited states of a BP molecule comprise higher-energy Soret states and lower-energy Q states. Here, we considered the Q states for analyzing the exciton diffusion length. The TDDFT calculations of the degenerate Q states of BP predicted a transition dipole moment of 1.46 a.u. and an excitation energy of 2.1 eV.

The exciton coupling, V_{XT}, of the parallel H-aggregate is relatively large compared with other directions (Table 2.1). Because the charge transfer coupling of the parallel H-aggregate is large (Table 2.1), the electron exchange, i.e. Dexter transfer, would contribute significantly to the exciton coupling. As for the other directions, the electron exchange is small and thus the Coulomb coupling, i.e. Förster transfer, is dominant.

Figure 2.12c, d show the exciton mobility in the BP crystal, based on the same definition of the charge mobility. The lifetime of the excited state (Q state) of BP was estimated to be 50 ns based on 2.7. Consequently, the exciton diffusion length, L_D, during 50 ns was estimated to be few hundred nm based on 2.12 (Fig. 2.12e, f). The estimated L_D of the singlet exciton is generally larger than the experimentally

Table 2.1 Intermolecular electronic coupling for the hole transfers, V_{hole} (eV), and the exciton transfers, V_{XT} (eV), between neighboring BP molecules calculated by DFT and TDDFT

BP pair	V_{hole} (eV)	V_{XT} (eV)	R_{ij} (Å)
Herring-1	0.0038	0.0059	9.426
Herring-2	0.0072	0.0090	11.210
H-parallel	0.1207	0.0155	6.591
J-parallel	0.0036	0.0083	12.405

R_{ij} (Å) is the center-to-center distance

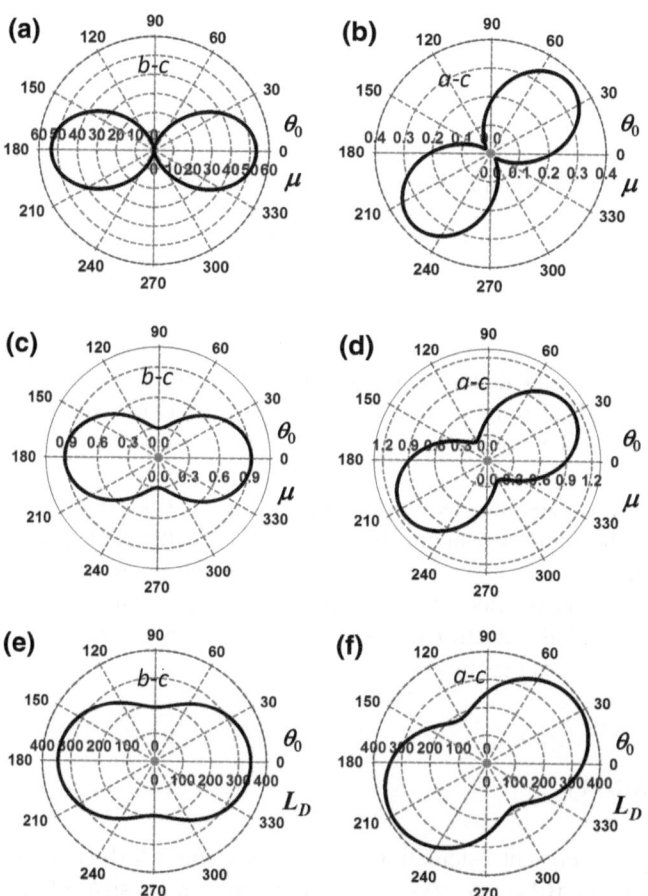

Fig. 2.12 Angle-dependent hole mobility, μ (cm^2/Vs), on the **a** b–c and **b** a–c planes in the BP crystal, where θ_0 is the angle from the b-axis and a-axis, respectively (see Fig. 2.1). The reorganization energy, λ for the hole transfer is 0.052 eV. Angle-dependent exciton mobility, μ (cm^2/Vs), on the **c** b–c and **d** a–c planes, where θ_0 is the angle from the b-axis and a-axis, respectively. Exciton diffusion length, L_D (nm), on the **e** b–c and **f** a–c planes, where the exciton life time is set to 50 ns. The reorganization energy, λ for the exciton transfer is 0.025 eV

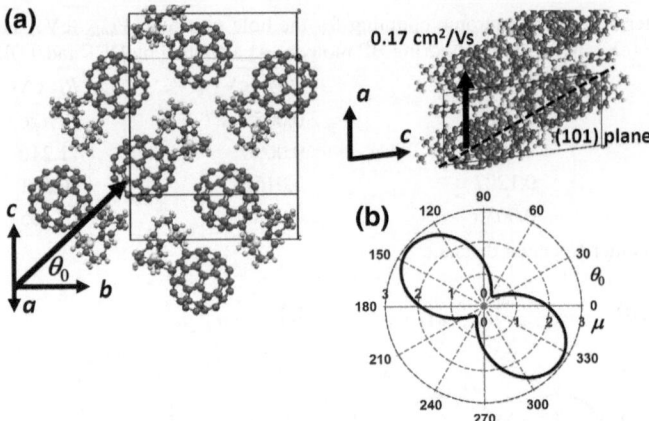

Fig. 2.13 **a** Crystal structure of SIMEF. The *box* indicates the unit cell. The cell parameters are as follows: $a = 10.360$, $b = 19.020$, $c = 22.665$ (Å), $\alpha = 90.0$, $\beta = 97.137$, $\gamma = 90.0$ (degree). The *dashed line* indicates the (101) plane. The *left panel* is a *top view* of the (101) plane. **b** Angle-dependent electron mobility, μ (cm²/Vs), on the (101) plane. The reorganization energy, λ for the electron transfer is 0.20 eV

reported values for related molecular condensates [48, 49]. This would be because the lifetime and the diffusion length of excitons are restricted by the grain boundaries in realistic molecular condensates, while the present model assumes ideal single crystals.

We further analyzed the electron mobility in the SIMEF crystal. For illustration, we estimated the angle-dependent electron mobility on the (101) plane (Fig. 2.13b), as well as the mobility along a-axis (Fig. 2.13a). The intermolecular coupling of the electron transfer is affected by the steric hindrance of the functional groups in SIMEF. As a result, the electron transport is efficient only along a direction that does not cross the functional groups (Fig. 2.13).

2.3.3 Summary

In summary, the present calculations predict an exciton diffusion length of a few hundred nm in the BP single crystal. Given that the typical sizes of the BP columns are within few tens of nm, the photo-generated exciton can easily reach the BP/SIMEF interface within the exciton lifetime. In the actual devices, the BP columns might be polycrystalline in which the domain boundary decreases the exciton diffusion length. Nevertheless, the loss of internal quantum efficiency during the exciton diffusion in the BP columns is expected to be rather small. It would remain to be seen how domain boundaries of polycrystals affect the exciton and charge transports.

2.4 Organic Light-Emitting Transistors

Organic light emitting transistors (OLET) [62–67] are of great interest owing to their potential application for optoelectronic devices including electrically driven organic lasers. Organic materials have various advantages for optoelectronic devices such as the tunability of the wavelength and polarization by chemical modifications and molecular orientations. The electrically driven photoluminescence of ambipolar OLETs is realized through the injection of electron and hole from the electrodes, the charge transports in the organic layer, exciton formation, and radiative deexcitation [62–67]. While the electrically driven organic lasers have not yet been developed successfully, amplified spontaneous emissions (ASE) have been achieved by optical pumping of organic single crystals [68–74].

The optoelectronic properties of thiophene-based π-conjugation molecules have been investigated extensively as standard organic semiconductors for OLET [62–67]. The single crystals of some thiophene oligomers [70, 71] and thiophene-phenylene co-oligomers [72, 74], e.g., 2,5-bis(4-biphenylyl) Bithiophene (BP2T) (Fig. 2.14) [72], exhibit ASE by optical pumping.

The stacking configurations of π-conjugation molecules in organic crystals can be classified into H-aggregate and J-aggregate. Closely packed H-aggregates with a strong electronic coupling are generally advantageous for the charge mobility, but are not favorable for the photoluminescence efficiency, since in general the exciton in H-aggregates decays rapidly to the dark state [75]. The lowest exciton state of J-aggregates is optically bright, but the charge mobility along J-aggregates is generally small. That is, the trade-off between the charge mobility and the photoluminescence efficiency is a crucial problem for realizing the optoelectronic devices such as the electrically driven organic lasers.

Recently, furan-incorporated π-conjugation molecules, which can exhibit high charge mobility and efficient photoluminescence [76, 77], have attracted increasing

Fig. 2.14 Crystal structure of BP2T and BPFT

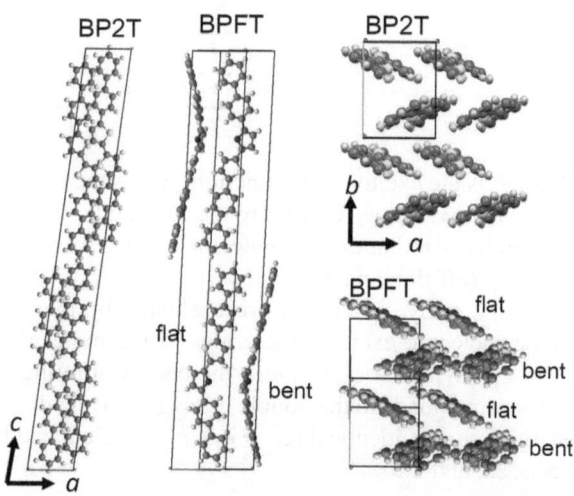

attention as organic semiconductors for optoelectronic devices. Oniwa et al. [78] have recently synthesized a novel thiophene-furan-phenylene co-oligomer, namely 2-(4-biphenyl)-5- [5-(4-biphenyl)-2- thienyl]furan (BPFT) (Fig. 2.14), aiming to improve the optoelectronic properties for OLET. The molecular structure of BPFT is similar to BP2T except that one of the thiophene rings of BP2T is replaced with furan. It is remarkable that the BPFT crystal exhibits much higher photoluminescence efficiency [78] and ASE intensity than the BP2T crystal, while the charge mobilities of the BPFT and BP2T crystals are of the same order of magnitude.

The X-ray crystallography revealed that the BPFT crystal takes an unexpected packing structure in which half of the molecules bend the π-conjugation plane (Fig. 2.14), while the BP2T crystal takes an ordinary herringbone packing structure. In view of the material design for OLET, it is curious how the bending of π-conjugation plane in the organic crystal improves the optoelectronic properties.

We have theoretically analyzed the photoluminescence efficiency of the BP2T and BPFT crystals based on the Frenkel exciton model and the vibronic coupling analysis parameterized by the time-dependent DFT (TDDFT) calculations [79]. Our analysis indicates that the high photoluminescence efficiency of the BPFT crystal originates from the symmetry breaking of the H-aggregate due to the bent molecular structure.

2.4.1 Methods

The exciton in molecular crystals is generally described by the Frenkel exciton model. Here, an exciton delocalized over molecules, i.e. coherent exciton, is described by a linear combination of the excited states of single molecules as follows:

$$
\begin{aligned}
H\Psi &= E_{ex}\Psi \\
\Psi &= \sum_i C_i \psi_i \\
H &= \sum_i \epsilon_i |i\rangle\langle i| + \sum_{i,j} h_{ij} |i\rangle\langle j|
\end{aligned}
\tag{2.13}
$$

Here, Ψ is the exciton wavefunction, E_{ex} is the exciton energy, and H is the Hamiltonian matrix. C_i, ψ_i, and ϵ_i are the amplitude, wavefunction, and site energy of the exciton localized at the ith molecule, respectively. h_{ij} is the intermolecular exciton coupling (off-diagonal term).

In this study, ϵ_i and h_{ij} are evaluated by the TDDFT calculations. When the single molecule excitation energies in the molecular aggregate are identical, i.e., $\epsilon_i = \epsilon_j$, (symmetric excimer), the energy splitting of the dark and bright exciton states corresponds to the double of h_{ij} (i.e., Dayvdov splitting). When the excitation energies are not identical ($\epsilon_i \neq \epsilon_j$), h_{ij} is calculated by the quasi-diabatization (see Sects. 2.2 and 2.3).

The transition dipole moment, d_{ex}, and the oscillator strength, f_{ex}, of a delocalized exciton are evaluated as follows (in a.u.):

$$d_{ex} = \langle \Psi_g | r | \Psi_i \rangle = \sum_i C_i d_i$$

$$f_{ex} = \frac{2}{3} E_{ex} |d_{ex}|^2 \tag{2.14}$$

where ψ_g is the ground state wavefunction, r is the dipole operator, and d_i is the transition dipole moment of the ith molecule. In general, the lower and higher excitons of H-aggregate are optically dark and bright, respectively, where the transition dipole of the dark state cancels out as the signs of C_i are opposite. In reverse, the lower and higher excitons of J-aggregate are bright and dark, respectively. That is, the sign of h_{ij} of H- and J-aggregates are positive and negative, respectively.

The energies and the transition dipole moments of the exciton states are calculated by diagonalizing the Hamiltonian matrix (2.6) under three-dimensional periodic boundary conditions, where we consider $8 \times 8 \times 4$ (256) molecules in the unit cell. The extent to which the exciton is localized in the molecular crystal is determined by the disorder of lattice that induces inhomogeneity of h_{ij}. For analyzing the optical spectra, the lattice disorder due to thermal fluctuations is taken into account based on the Monte Carlo sampling at 300 K [79]. The light emission of organic crystals is assumed to occur after the exciton decays to the lower states. To obtain the emission spectra under thermal equilibrium, the oscillator strength is weighted by the Boltzmann factor as follows:

$$\text{Intensity}(E_{ex}) = \frac{\exp(-E_{ex}/kT)}{Z} f_{ex}(E_{ex}) \tag{2.15}$$

where Z is the partition function.

The emission spectrum accounting for the vibronic coupling, i.e., the Franck-Condon factor, is calculated as follows. We calculate the dynamics of vibrational wave-packet on the ground state potential starting from the equilibrium position of the excited state, where the frequency and vibronic coupling of the normal modes are evaluated by the DFT calculations with the harmonic approximation. Then, the vibronic spectrum is obtained by the Fourier transform of the auto-correlation function of wave-packet dynamics.

2.4.2 Emission Spectra of Organic Crystals

We calculate the excitation energy (ϵ_i), exciton coupling (h_{ij}), and transition dipole moment (d_i) by TDDFT. The oscillator strength of the BP2T and BPFT single molecules are similar; therefore, the high photoluminescence efficiency of the BPFT crystal cannot be explained by the properties of single molecule. The excitation energy of the bent BPFT is larger than the flat one by 0.053 eV.

Fig. 2.15 Transition dipole moment of the lower exciton state (d_{low}) for the H-aggregates in the BP2T and BPFT crystals, where *arrows* illustrate the single molecule transition dipoles

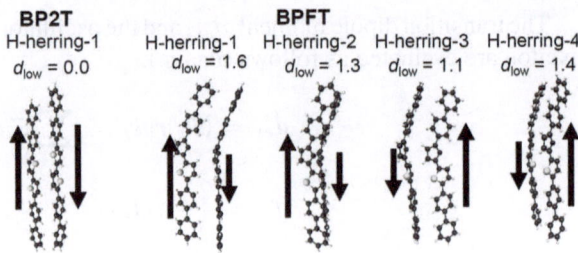

The h_{ij} of H-aggregates (a–b plane) are found to be larger than those of J-aggregates (c-axis) [79]. In such cases, the lowest exciton state in the crystal generally becomes optically dark. Figure 2.15 summarizes the transition dipole moment, d, of the H-aggregates of BP2T and BPFT.

While the transition dipole moment cancels out in the dark state of the BP2T H-aggregates, the transition dipole moment does not completely cancel out in the dark state of the BPFT H-aggregate consisting of the bent and flat molecules (Fig. 2.15). In the dark state, the exciton population at the flat BPFT molecule is somewhat larger than at the bent one, because of the difference in the excitation energy of the bent and flat molecules. This symmetry-breaking results in non-zero transition dipole moment of the dark state of the BPFT H-aggregate.

We calculate the spectra of the exciton states in the BP2T and BPFT crystals (Fig. 2.16), based on the Frenkel exciton model parameterized by TDDFT. The long wavelength region of the BP2T spectrum consists basically of the dark states, since the exciton couplings of the H-aggregates (a–b plane) are stronger than those of the J-aggregates (c axis). The transition dipole cancels out in the dark states of the completely ordered BP2T crystal. The lattice fluctuations induce inhomogeneity

Fig. 2.16 a Vibronic emission spectra of the BP2T and BPFT single molecules, together with the diagram of vibronic states. **b** Electronic emission spectra of the organic crystals considering the Boltzmann distribution of the exciton states at 300 K. **c** Emission spectra of the organic crystals accounting for the vibronic coupling, where the electronic spectra of the crystals are multiplied by the vibronic spectra of the single molecules

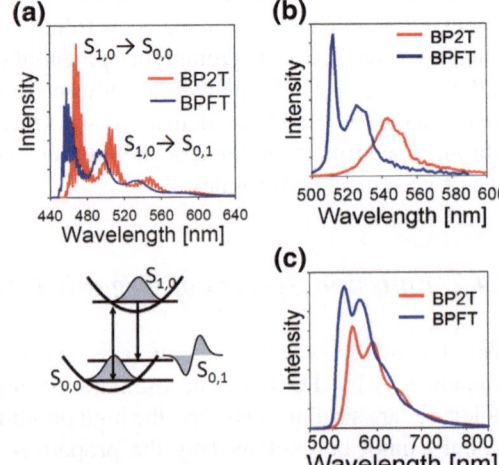

of the intermolecular distances and exciton couplings, thereby increasing the net transition dipole of the dark state.

A noticeable peak appears in the long wavelength region of the calculated BPFT spectrum, while such a peak is not observed for BP2T (Fig. 2.16b). In this exciton state, the direction of transition dipole is opposite at the bent and flat BPFT molecules, but the total transition dipole does not cancel out even if the lattice is completely ordered. This is because the exciton population, $|C_i|^2$, at the flat molecules tends to be larger than at the bent molecules owing to the difference in the excitation energy, ϵ_i. Thus, the exciton in the BPFT crystal is not quenched by the relaxation to the dark states of the H-aggregates, in contrast to the BP2T crystal that takes ordinary herringbone packing. This would be a reason why the photoluminescence efficiency of the BPFT crystal is higher than BP2T, even though the transition dipole moments of the BP2T and BPFT single molecules are similar.

We also analyze the influence of the vibronic coupling, i.e., the Franck-Condon factor, on the emission spectra. Figure 2.16a shows the vibronic emission spectra of the BP2T and BPFT single molecules calculated by the wave-packet dynamics parameterized by the DFT calculations. The shortest wavelength peak corresponds to the transition from the vibrational ground state of the electronic excited state ($S_{1,0}$) to the vibrational ground state of the electronic ground state ($S_{0,0}$). The second peak corresponds to the transition from $S_{1,0}$ to the vibrational excited state of the electronic ground state ($S_{0,1}$).

The emission spectra of the BP2T and BPFT crystals (Fig. 2.16b) are multiplied by the vibronic spectra (Fig. 2.16a) in order to obtain realistic emission spectra (Fig. 2.16c). The emission maxima of the calculated spectra are in good agreement with those of experimental spectra [78, 79]. The calculated emission intensity of the BPFT crystal is higher than that of the BP2T crystal consistent with the experimental photoluminescence efficiency. The high photoluminescence efficiency of the BPFT crystal is considered to originate from the symmetry-breaking of the H-aggregates, which increases the transition dipole moments of the dark states.

2.4.3 Summary

Our theoretical analysis revealed fundamental mechanisms underlying the optoelectronic properties of the phenylene-thiophene-furan co-oligomer crystals. The asymmetric herringbone structure consisting of the bent and flat BPFT molecules originates in the remarkable optical properties. The transition dipole moment does not cancel out in the dark state of the H-aggregate consisting of the bent and flat BPFT molecules. As a result, the BPFT crystal can exhibit an efficient photoluminescence, even if the bright exciton decays rapidly to the dark states. This is contrary to the conventional picture that the exciton is quenched in closely stacked H-aggregates.

The bending of π-conjugation molecules that breaks the symmetry of the H-aggregate is advantageous for increasing the charge mobility and the photoluminescence efficiency simultaneously, since the exciton relaxation to the dark states does not prevent the light emission, even if the intermolecular electronic coupling is increased by decreasing the $\pi-\pi$ stacking distance. This feature provides a guiding principle for solving the trade-off between the charge mobility and the photoluminescence efficiency for OLET.

References

1. L. Dou, J. You, Z. Hong, Z. Xu, G. Li, R.A. Street, Y. Yang, Adv. Mater. **25**, 6642 (2013)
2. C. Deibel, T. Strobel, V. Dyakonov, Adv. Mater. **22**, 4097 (2010)
3. P.W.M. Blom, V.D. Mihailetchi, L.J.A. Koster, D.E. Markov, Adv. Mater. **19**, 1551 (2007)
4. I.W. Hwang, C. Soci, D. Moses, Z. Zhu, D. Waller, R. Gaudiana, C.J. Brabec, A.J. Heeger, Adv. Mater. **19**, 2307 (2007)
5. S.R. Cowan, N. Banerji, W.L. Leong, A.J. Heeger, Adv. Funct. Mater. **22**, 1116 (2012)
6. C.J. Brabec, G. Zerza, G. Cerullo, S. De Silvestri, S. Luzzati, J.C. Hummelen, S. Sariciftci, Chem. Phys. Lett. **340**, 232 (2001)
7. R.D. Pensack, J.B. Asbury, J. Am. Chem. Soc. **131**, 15986 (2009)
8. J. Lee, K. Vandewal, S.R. Yost, M.E. Bahlke, L. Goris, M.A. Baldo, J.V. Manca, T. Van Voorhis, J. Am. Chem. Soc. **132**, 11878 (2010)
9. J. Guo, H. Ohkita, H. Benten, S. Ito, J. Am. Chem. Soc. **132**, 6154 (2010)
10. T. Miura, M. Aikawa, Y. Kobori, J. Phys. Chem. Lett. **5**, 30 (2014)
11. A.A. Bakulin, A. Rao, V.G. Pavelyev, P.H.M. van Loosdrecht, M.S. Pshenichnikov, D. Niedzialek, J. Cornil, D. Beljonne, R.H. Friend, Science **335**, 1340 (2012)
12. S.D. Dimitrov, A.A. Bakulin, C.B. Nielsen, B.C. Schroeder, J. Du, H. Bronstein, I. McCulloch, R.H. Friend, J.R. Durrant, J. Am. Chem. Soc. **134**, 18189 (2012)
13. S. Gélinas, A. Rao, A. Kumar, S.L. Smith, A.W. Chin, J. Clark, T.S. van der Poll, G.C. Bazan, R.H. Friend, Science **343**, 512 (2014)
14. L.G. Kaake, D. Moses, A.J. Heeger, J. Phys. Chem. Lett. **4**, 2264 (2013)
15. A.E. Jailaubekov, A.P. Willard, J.R. Tritsch, W.-L. Chan, N. Sai, R. Gearba, L.G. Kaake, K.J. Williams, K. Leung, P.J. Rossky, X.-Y. Zhu, Nat. Mater. **12**, 66 (2013)
16. G. Grancini, M. Maiuri, D. Fazzi1, A. Petrozza1, H.-J. Egelhaaf, D. Brida, G. Cerullo, G. Lanzani. Nat. Mater. **12**, 29 (2013)
17. Y. Yuan, T.J. Reece, P. Sharma, S. Poddar, S. Ducharme, A. Gruverman, Y. Yang, J. Huang, Nat. Mater. **10**, 296 (2011)
18. P.K. Nayak, K.L. Narasimhan, D. Cahen, J. Phys. Chem. Lett. **4**, 1707 (2013)
19. T.J. Savenije, J.E. Kroeze, X. Yang, J. Loos, Adv. Funct. Mater. **15**, 1260 (2005)
20. Y. Kim, S. Cook, S.M. Tuladhar, S.A. Choulis, J. Nelson, J.R. Durrant, D.D.C. Bradley, M. Giles, I. McCulloch, C.-S. Ha, M. Ree, Nat. Mater. **5**, 197 (2006)
21. C. Deibel, T. Strobel, V. Dyakonov, Phys. Rev. Lett. **103**, 036402 (2009)
22. H. Tamura, E.R. Bittner, I. Burghardt, J. Chem. Phys. **127**, 034706 (2007)
23. H. Tamura, J.G.S. Ramon, E.R. Bittner, I. Burghardt, Phys. Rev. Lett. **100**, 107402 (2008)
24. H. Tamura, I. Burghardt, M. Tsukada, J. Phys. Chem. C **115**, 10205 (2011)
25. H. Tamura, R. Martinazzo, M. Ruckenbauer, I. Burghardt, J. Chem. Phys. **137**, 22A540 (2012)
26. H. Tamura, I. Burghardt, J. Phys. Chem. C **117**, 15020 (2013)
27. H. Tamura, I. Burghardt, J. Am. Chem. Soc. **135**, 16364 (2013)
28. D.P. McMahon, D.L. Cheung, A. Troisi, J. Phys. Chem. Lett. **2**, 2737 (2011)
29. D. Caruso, A. Troisi, Proc. Natl. Acad. Sci. USA **109**, 13498 (2012)
30. S.R. Yost, L.-P. Wang, T.V. Voorhis, J. Phys. Chem. C. **115**, 14431 (2011)
31. S. Verlaak, D. Beljonne, D. Cheyns, C. Rolin, M. Linares, F. Castet, J. Cornil, P. Heremans, Adv. Funct. Mater. **19**, 3809 (2009)
32. B.P. Rand, D. Cheyns, K. Vasseur, N.C. Giebink, S. Mothy, Y. Yi, V. Coropceanu, D. Beljonne, J. Cornil, J.L. Brédas, J. Genoe, Adv. Funct. Mater. **22**, 2987 (2012)

33. E.R. Bittner, C. Silva, Nat. Commun. **5**, 3199 (2014)
34. R.A. Marcus, Rev. Mod. Phys. **65**, 599 (1993)
35. C.L. Braun, J. Chem. Phys. **80**, 4157 (1984)
36. H.D. Meyer, U. Manthe, L.S. Cederbaum, Chem. Phys. Lett. **165**, 73 (1990)
37. M.H. Beck, A. Jäckle, G.A. Worth, H.D. Meyer, Phys. Rep. **324**, 1 (2000)
38. Y. Towada, T. Tsuneda, S. Yanagisawa, Y. Yanai, K. Hirao, J. Chem. Phys. **120**, 8425 (2004)
39. S. Dag, L.W. Wang, J. Phys. Chem. B **114**, 5997 (2010)
40. H. Tamura, M. Tsukada, Phys. Rev. B **85**, 054301 (2012)
41. Y. Matsuo, Y. Sato, T. Niinomi, I. Soga, H. Tanaka, E. Nakamura, J. Am. Chem. Soc. **131**, 16048 (2009)
42. H. Tanaka, Y. Abe, Y. Matsuo, J. Kawai, I. Soga, Y. Sato, E. Nakamura, Adv. Mater. **24**, 3521 (2012)
43. B. Walker, C. Kim, T.-Q. Nguyen, Chem. Mater. **23**, 470 (2011)
44. M. Guide, X.-D. Dang, T.-Q. Nguyen, Adv. Mater. **23**, 2313 (2011)
45. Y. Matsuo, A. Iwashita, Y. Abe, C. Li, K. Matsuo, M. Hashiguchi, E. Nakamura, J. Am. Chem. Soc. **130**, 15429 (2008)
46. Y. Matsuo, J. Hatano, T. Kuwabara, K. Takahashi, Appl. Phys. Lett. **100**, 063303 (2012)
47. H. Tamura, Y. Matsuo, Chem. Phys. Lett. **598**, 81 (2014)
48. S.R. Scully, M.D. McGeheea, J. Appl. Phys. **100**, 034907 (2006)
49. R.R. Lunt, N.C. Giebink, A.A. Belak, J.B. Benziger, S.R. Forrest, J. Appl. Phys. **105**, 053711 (2009)
50. P. Irkhin, I. Biaggio, Phys. Rev. Lett. **107**, 017402 (2011)
51. V. Podzorov, E. Menard, A. Borissov, V. Kiryukhin, J.A. Rogers, M.E. Gershenson, Phys. Rev. Lett. **93**, 086602 (2004)
52. V. Podzorov, E. Menard, J.A. Rogers, M.E. Gershenson, Phys. Rev. Lett. **95**, 226601 (2005)
53. J. Takeya, M. Yamagishi, Y. Tominari, R. Hirahara, Y. Nakazawa, T. Nishikawa, T. Kawase, T. Shimoda, S. Ogawa, Appl. Phys. Lett. **90**, 102120 (2007)
54. J. Takeya, J. Kato, K. Hara, M. Yamagishi, R. Hirahara, K. Yamada, Y. Nakazawa, S. Ikehata, K. Tsukagoshi, Y. Aoyagi, T. Takenobu, Y. Iwasa, Phys. Rev. Lett. **98**, 196804 (2007)
55. T. Uemura, M. Yamagishi, J. Soeda, Y. Takatsuki, Y. Okada, Y. Nakazawa, J. Takeya, Phys. Rev. B **85**, 035313 (2012)
56. H. Sumi, J. Phys. Chem. B **103**, 252 (1999)
57. G.R. Fleming, G.D. Scholes, Nature **431**, 256 (2004)
58. W. Deng, W.A. Goddard III, J. Phys. Chem. B **108**, 8614 (2004)
59. S. Wen, A. Li, J. Song, W. Deng, K. Han, W.A. Goddard III, J. Phys. Chem. B **113**, 8813 (2009)
60. H. Tamura, M. Tsukada, H. Ishii, N. Kobayashi, K. Hirose, Phys. Rev. B **86**, 035208 (2012)
61. H. Tamura, M. Tsukada, H. Ishii, N. Kobayashi, K. Hirose, Phys. Rev. B **87**, 155305 (2013)
62. M.A. Muccini, Nat. Mater. **5**, 605 (2006)
63. J. Zaumseil, R.H. Friend, H. Sirringhaus, Nat. Mater. **5**, 69 (2006)
64. T. Takahashi, T. Takenobu, J. Takeya, Y. Iwasa, Adv. Funct. Mater. **17**, 1623 (2007)
65. T. Takenobu, S.Z. Bisri, T. Takahashi, M. Yahiro, C. Adachi, Y. Iwasa, Phys. Rev. Lett. **100**, 066601 (2008)
66. S.Z. Bisri, T. Takenobu, Y. Yomogida, H. Shimotani, T. Yamao, S. Hotta, Y. Iwasa, Adv. Funct. Mater. **19**, 1728 (2009)
67. Y. Wang, D. Liu, S. Ikeda, R. Kumashiro, R. Nouch, Y. Xu, H. Shang, Y. Ma, K. Tanigaki, Appl. Phys. Lett. **97**, 033305 (2010)
68. I.D.W. Samuel, G.A. Turnbull, Chem. Rev. **107**, 1272 (2007)
69. P.A. Losio, C. Hunziker, P. Gunter, Appl. Phys. Lett. **90**, 241103 (2007)
70. D. Fichou, S. Delysse, J. Nunzi, Adv. Mater. **9**, 1178 (1997)
71. M. Polo, A. Camposeo, S. Tavazzi, L. Raimondo, P. Spearman, A. Papagni, R. Cingolani, D. Pisignano, Appl. Phys. Lett. **92**, 083311 (2008)
72. M. Ichikawa, R. Hibino, M. Inoue, T. Haritani, S. Hotta, T. Koyama, Y. Taniguchi, Adv. Mater. **15**, 213 (2003)

73. M. Ichikawa, K. Nakamura, M. Inoue, H. Mishima, T. Haritani, R. Hibino, T. Koyama, Y. Taniguchi, Appl. Phys. Lett. **87**, 221113 (2005)
74. A. Saeki, S. Seki, Y. Shimizu, T. Yamao, S. Hotta, J. Chem. Phys. **132**, 134509 (2010)
75. H. Marciniak, M. Fiebig, M. Huth, S. Schiefer, B. Nickel, F. Selmaier, S. Lochbrunner, Phys. Rev. Lett. **99**, 176402 (2007)
76. O. Gidron, A. Dadvand, Y. Sheynin, M. Bendikov, D.F. Perepichka, Chem. Commun. **47**, 1976 (2011)
77. C. Mitsui, J. Soeda, K. Miwa, H. Tsuji, J. Takeya, E. Nakamura, J. Am. Chem. Soc. **134**, 5448 (2012)
78. K. Oniwa, T. Kanagasekaran, T. Jin, Md. Akhtaruzzaman, Y. Yamamoto, H. Tamura, I. Hamada, H. Shimotani, N. Asao, S. Ikeda, K. Tanigaki, J. Mater. Chem. C **1**, 4163 (2013)
79. H. Tamura, I. Hamada, H. Shang, K. Oniwa, M. Akhtaruzzaman, T. Jin, N. Asao, Y. Yamamoto, K. Thangavel, H. Shimotani, S. Ikeda, K. Tanigaki. J. Phys. Chem. C **117**, 8072 (2013)
80. K. Sen, R. Crespo-Otero, O. Weingart, W. Thiel, M. Barbatti, J. Chem. Theory Comput. **9**, 533 (2013)

Chapter 3
Laser Spectroscopy Using Topological Light Beams

Yasunori Toda and Ryuji Morita

Abstract Simplifications of systems are important for understanding their universal and/or intrinsic properties. Topology is one of the key concepts of such procedures, which focuses simply on the connectivity of the system to clarify the essential aspects of the geometric structures. To date, this concept has been extended to the field of physics, especially to the condensed matter physics and materials science. For example, topological defects have been observed in various materials, such as liquid crystals, superconductors, and electron gas systems in semiconductors. More recently, photoexcitations to some specific materials (mainly semiconductor nanostructures) have also revealed formation of topological defects. On the other hand, optical field itself includes a topological character, which has been well known as "optical or polarization vortex", "twisted light", and "helical or Laguerre-Gauss light" beams. These topological light beams exhibit spiral (spatial) variations of the phase (polarization) producing phase (polarization) singularities on the wavefront, which can be regarded as topological defects (screw dislocations). Therefore, we have possibilities to evaluate the topological aspects of material system on the interactions with topological light beams. However, the question arises: does it make any sense to apply the topological concept to the laser spectroscopy? The purpose of this chapter is to answer the question by introducing our recent research on this topic. Experimental results will be presented and discussed in terms of topological order of electrons. The chapter begins with the basics of topological light beams together with their important properties for laser spectroscopy (Sect. 3.1). Both the historical background and the overview of applications will also be introduced. In Sects. 3.2 and 3.3, we present several techniques for generating and evaluating the topological light beams, which are useful and needed in the experimental sections. We also discuss their advantages and drawbacks. In Sects. 3.4 and 3.5, we present our experimental results on

Y. Toda (✉) · R. Morita
Department of Applied Physics, Hokkaido University, Nishi-8,
Sapporo 060-8628, Japan

Core Research for Evolutionary Science and Technology (CREST), Japan Science and
Technology Agency (JST), Kawaguchi, Japan
e-mail: toda@eng.hokudai.ac.jp

R. Morita
e-mail: morita@eng.hokudai.ac.jp

© Springer International Publishing Switzerland 2015
M. Ohtsu and T. Yatsui (eds.), *Progress in Nanophotonics 3*,
Nano-Optics and Nanophotonics, DOI 10.1007/978-3-319-11602-0_3

nonlinear four-wave-mixing and pump-probe reflection spectroscopy, respectively. In both cases, the spatial characteristics of the topological light beams allow us to investigate unique properties associated with topologically-ordered electrons in materials. Section 3.6 summarizes this chapter.

3.1 Basics of Topological Light Beams

3.1.1 Introduction

In most of the chapter, we treat the so-called optical vortices as a topological light beam (Fig. 3.1a). They are also called optical phase singularities or wave disloca-tions. We also consider the polarization vortices characterized by the spatial variation of the polarization field (Fig. 3.1b). A remarkable feature of connecting the optical (polarization) vortices to "topology" is the presence of the phase (polarization) sin-gularities, which has attracted attention recently. However, topological singularity of light has a long history dating back to the early 19th century, which will be cited later. In the optical vortex, the name of vortex comes from the phase gradient around the singularity on the wavefront characterized by $\exp[i\ell\phi]$, where ϕ is the azimuthal angle and ℓ is any integer value determined by the winding number of the $\pm 2\pi$ phase change [35]. Since the propagation of the optical vortex produces a spiral (helical) wavefront, we sometimes call spiral beam, helical beam, or twisted light. On the spiral wavefront, the energy flow (Poynting vector) rotates around the nodal line (trajectory of the phase singularity), resulting in a topological charge (orbital angular momentum) given by $\ell(\ell\hbar)$ [1]. The polarization vortex is another class of topological light beam, where the axisymmetrical polarization produces the vortex. In this subsection, we briefly overview these characteristics one by one.

First we associate the optical vortex with transverse modes of the light beam, which is known as Laguerre-Gauss (LG) mode [9, 20, 30]. The use of LG is appropriate when we focus on the individual transverse modes, especially on the laser cavity modes. Since the birth of the laser, the research on phase sin-gularities of the transverse modes were developed rapidly in terms of nonlin-

(a)   **(b)**

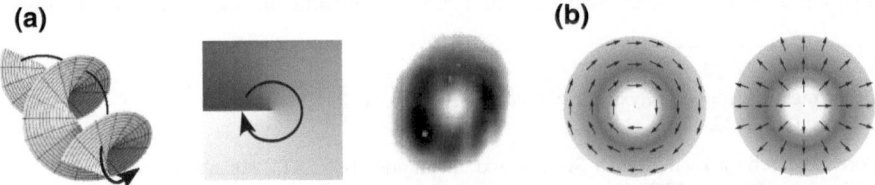

Fig. 3.1 Schematic illustrations of topological light beams. **a** Helical phase front (*left*), cross-sectional phase (*center*), and intensity (*right*) of an optical vortex beam with $\ell = 1$. **b** Azimuthally- (*left*) and radially- (*right*) polarized vortex beams

ear interactions of topological light with materials, including gases, liquids, and solids. Because of a wide class of nonlinear materials, the research have also made significant advances in understanding the topological physics of material science. Moreover, in the cavity modes, the interaction and competition among the transverse modes result in a rich variety of pattern formation of vortices (phase singularities), which can be linked to the topological defects in hydrodynamics, superconductivity, and superfluidity. The transverse dynamics of the phase singularities in cavity devices has now developed to the concept of photon fluids and connected further to the topological physics [12, 24].

In the context of vortex dynamics (or connectivities of nodal lines of phase singularities), propagations of optical vortices provide important aspects of topology [23, 36, 38]. In the propagation, phase singularities produce nodal lines as a result of undefined phase and these lines show various linkage and knottedness of vortex filaments in nonlinear optical media [17, 29]. Since the phase singularities are observed as dark points (null intensities) within the beam, one can easily obtain the trajectory of the phase singularities in successive beam cross-sections and evaluate their topological dynamics. Another advantage of the dark point derives from the property of spatial invariance, where the dark point can be kept in the focal point even when focused by lenses with high numerical aperture and its size is ideally beyond the diffraction limit. As a result, pure LG beam whose singularity is located on the center of the beam cross-section can be used in the nanofabrication, superresolution imaging, and so on [10, 19, 22].

Next we consider the Poynting vector of optical vortices. Because of the spiral wavefront, Poynting vector of the vortex beam has an azimuthal energy flow around the vortices, giving rise to a rotational torque within the beam cross-section. This means that the beam with a phase gradient $\exp[i\ell\phi]$ carries orbital angular momentum (OAM) of $\ell\hbar$ per photon, which has been pointed out by Allen et al. in 1992 [1]. Thanks to such an essential definition, the research on optical vortices has been extensively developed, not only for fundamental physics but also for technological applications, such as optical trapping (including BEC), manipulation, laser fabrication, information processing [17].

On the other hand, polarization is another degree of freedom of light field, which corresponds to the spin angular momentum (SAM) of photon. Because of the simple relationship with electron spin, the optical polarization is more general in the laser spectroscopy than the optical OAM. In the topological aspect, one can easily recognize that the spatial variations of the polarization field produce the singularities, the first observation of which was given by Arago in the early 19th century [6]. As is similar to the LG beam, the polarization vortex beam can be observed in the transverse mode of the laser, which is so-called "vector beam" [34, 37]. In several parts of the chapter, we employ the vector beams with radial and azimuthal polarizations (Fig. 3.1b). These axisymmetrically-polarized beams have attracted much attention recently because they are known to produce longitudinal electric and magnetic fields when tightly focused by a high numerical aperture lens [54].

As a summary of this introduction, the concept of topological light beams have a long history, but became widespread during the past decade. The physics and potential applications achieved by the topological light beams are now very broad ranging from nanophotonics to bioengineering, and astrophysics. Among them, we will focus on the applications to the laser spectroscopy.

3.1.2 Optical Vortex or Laguerre-Gauss Beam

For some basic physical concepts and mathematical descriptions of the optical vortex, we begin with the wave equation in vacuo. In free-space, the electric field vector of light is given by

$$\nabla^2 \mathbf{E} - \frac{1}{c^2} \frac{\partial^2 \mathbf{E}}{\partial t^2} = 0, \tag{3.1}$$

where c is the speed of light. Assuming a monochromatic field (or superposition of monochromatic fields for the optical pulse), we can separate the time dependence of the field given by $\exp[-i\omega t]$, and obtain the vector Helmholtz equation $(\nabla^2 + k^2) \mathbf{E} = 0$ (the propagation constant $k = \omega/c = 2\pi/\lambda$, where ω and λ are the angular frequency and the wavelength of light, respectively), or scalar Helmholtz equation $(\nabla^2 + k^2) E = 0$ if neglecting the polarization. For the spatial distribution of the beam cross-section $(x, y$ plane), the paraxial approximation is applied to the Helmholtz equation, which is represented by

$$\frac{\partial^2 E}{\partial x^2} + \frac{\partial^2 E}{\partial y^2}^2 + 2ik\frac{\partial E}{\partial z} = 0, \text{ or } \left(\nabla_T^2 + 2ik\frac{\partial}{\partial z}\right) E = 0, \tag{3.2}$$

where z is the propagation axis. The solutions of (3.2) can be described in terms of the orthogonal field basis functions. Here we focus on two sets of basis functions; Hermite-Gauss (HG) and Laguerre-Gauss (LG) beams, which form complete sets of orthogonal functions in the Cartesian (x, y, z) and cylindrical (r, ϕ, z) coordinates, respectively. Using $E(\mathbf{r}, z) = u(\mathbf{r}, z) \exp[i(kz - \omega t)]$, the transverse components of the beams are given by,

$$
\begin{aligned}
\mathbf{HG_n^m} : \; u_{m,n}(x, y, z) = {} & H_m\left[\frac{\sqrt{2}x}{w(z)}\right] \cdot H_n\left[\frac{\sqrt{2}y}{w(z)}\right] \\
& \cdot \frac{w_0}{w(z)} \cdot \exp\left[-\frac{x^2 + y^2}{w^2(z)}\right] \\
& \cdot \exp\left[-i(1 + m + n)\tan^{-1}\left(\frac{z}{z_R}\right)\right] \\
& \cdot \exp\left[-i\frac{k(x^2 + y^2)}{2R(z)}\right],
\end{aligned}
\tag{3.3}
$$

Table 3.1 Parameters and functions for (3.3) and (3.4)

Power density radius	$w(z) = w_0\sqrt{1 + z^2/z_R^2}$
Beam waist radius defined at $z = 0$	$w_0 = w(0)$
Radius for curvature of wavefront	$R(z) = \left(z^2 + z_R^2\right)/z$
Rayleigh range	$z_R = kw_0^2/2$

$$H_n(u) = (-1)^n\, e^{u^2}\, \frac{d^n}{du^n} e^{-u^2} = n! \sum_{m=0}^{\lfloor n/2 \rfloor} \frac{(-1)^m}{m!\,(n-2m)!}\,(2u)^{n-2m}$$

$$L_p^{|\ell|}(u) = \frac{u^{-|\ell|}e^u}{p!}\,\frac{d^p}{du^p}\left(e^{-u}u^{p+|\ell|}\right) = \sum_{m=0}^{p}(-1)^m \binom{p+|\ell|}{p-m}\frac{u^m}{m!}$$

and

$$\mathbf{LG}_\ell^\mathbf{p}: \quad u_{\ell,p}(r,\phi,z) = \sqrt{\frac{2p!}{\pi\,(p+|\ell|)!}}\left(\frac{\sqrt{2}r}{w(z)}\right)^{|\ell|}\cdot L_{|\ell|}^{p}\left[\frac{2r^2}{w(z)^2}\right]$$

$$\cdot\frac{w_0}{w(z)}\cdot\exp\left[-\frac{r^2}{w^2(z)}\right]\cdot\exp\left[i\ell\phi\right]$$

$$\cdot\exp\left[-i\,(1+2p+|\ell|)\tan^{-1}\left(\frac{z}{z_R}\right)\right]$$

$$\cdot\exp\left[-i\frac{kr^2}{2R(z)}\right], \tag{3.4}$$

where $R(z)$, $w(z)$, z_R, and w_0, and Hermite- and Laguerre-polynomials are given in Table 3.1.

In both cases, the lowest-order modes have a Gaussian profile, called TEM$_{00}$ mode that is the fundamental cavity mode of laser. The higher-order modes are also Gaussian-like profile, but is modified by the Hermite-polynomial H_m^n ($n, m \geq 0$) and Laguerre-polynomial $L_{|\ell|}^{p}$ ($p \geq 0$, ℓ is any integer) as shown in Fig. 3.2. In the LG cases with $\ell \neq 0$, the beam profiles are characterized by a null intensity at the center, which originates from $\exp[i\ell\phi]$ in (3.4). We also note that the complete set of orthogonal functions has crucial consequences for the vortex beams; (1) HG and LG modes are related with each other by simple transformations, (2) the optical vortices can be expressed as a series expansion of these functions, both of which will be important for the generation and evaluation of the vortex beams in Sects. 3.2 and 3.3. We note that, in most parts of the chapter, we will focus only on the azimuthal index ℓ (i.e. topological charge) in the LG beam and assume the radial index $p = 0$.

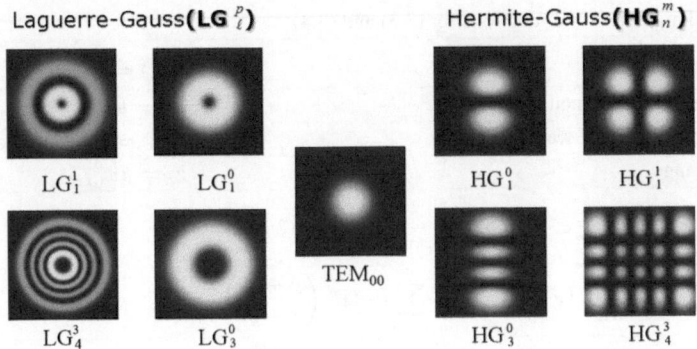

Fig. 3.2 Typical examples of Laguerre-Gauss (LG) and Hermite-Gaussian (HG) modes. In both cases, the fundamental mode is a TEM$_{00}$ mode

3.1.3 Orbital Angular Momentum (OAM) of LG Beam

The relationship between OAM and LG beams discovered by Allen et al. had a great impact on the development of research in this area [1]. Here we briefly summarize the OAM of the LG by introducing the Poynting vector $\mathbf{S} = \mathbf{E} \times \mathbf{H}$, where $\mathbf{E}$ and $\mathbf{H}$ are electric and magnetic fields of light, respectively. Since the energy flow $\mathbf{S}$ is parallel to the momentum flow $\mathbf{p} = \left(1/c^2\right)\mathbf{S}$ and is normal to the wavefront, the spiral wavefronts of the LG beam result in an azimuthal momentum component within the beam cross-section and hence an OAM in the propagation direction. As a simple definition, the OAM of a photon is given by $\ell\hbar$, where ℓ is denoted by the azimuthal mode index of the LG beam. In order to clarify this relationship, we present several formulations under Lorenz gauge condition. Using vector potential $\mathbf{A}$ and scalar potential ϕ, the electric field ($\mathbf{E}$) and magnetic field ($\mathbf{B}$) are described by

$$\mathbf{E} = -\frac{\partial \mathbf{A}}{\partial t} - \nabla\phi, \quad \mathbf{B} = \nabla \times \mathbf{A}. \tag{3.5}$$

For the propagation of the optical field in the cylindrical coordinate, the vector potential $\mathbf{A}$ can be written as

$$\mathbf{A} = \mathbf{n}u\left(r, \phi, z\right)\exp\left[i(kz - \omega t)\right], \tag{3.6}$$

where $\mathbf{n}$ represents the polarization. Assuming the paraxial beam propagation, we obtain

$$\mathbf{E} = i\omega\left(\mathbf{n}u + \mathbf{z}\frac{i}{k}\left(\mathbf{n}\cdot\nabla_T\right)u\right)\exp\left[ikz\right],$$

$$\mathbf{B} = ik\left(\mathbf{z}\times\mathbf{n}u + \frac{i}{k}\mathbf{n}\times\nabla_T u\right)\exp\left[ikz\right].$$

Here we neglect the time-dependence. The Poynting vector is then given by

$$S = \frac{1}{2\mu_0} \left(\mathbf{E}^* \times \mathbf{B} + \mathbf{B} \times \mathbf{E}^* \right) = i\omega \frac{1}{2\mu_0} \left(u^* \nabla_T u - u \nabla_T u^* \right) + \frac{\omega k}{\mu_0} |u|^2 \, \mathbf{e}_z$$
$$+ i\omega \frac{\varepsilon_0}{4} \left(\mathbf{n} \times \mathbf{n}^* - \mathbf{n}^* \times \mathbf{n} \right) \times \nabla_T |u|^2 , \tag{3.7}$$

where ε_0 and μ_0 are magnetic permeability and dielectric permittivity, respectively. Obviously $\nabla_T u$ in the first term represents the azimuthal component of the energy flow within the beam cross-section that is derived from the spatial phase gradient such as $\exp[i\ell\phi]$ for LG beams. The second term is the energy flow of light in the propagation direction. The last term is associated with the optical polarization and thus with the SAM of light, which becomes more apparent by the momentum density flow:

$$\mathbf{p} = i\omega \frac{\varepsilon_0}{2} \left(u^* \nabla_T u - u \nabla_T u^* \right) + \omega k \varepsilon_0 |u|^2 \mathbf{e}_z + \omega \sigma_z \frac{\varepsilon_0}{2} \frac{\partial |u|^2}{\partial r} \mathbf{e}_\phi. \tag{3.8}$$

Here $\sigma_z = \pm 1$ corresponds to the circularly polarized optical field. From (3.8), we can derive the angular momentum density of $\mathbf{j} = \mathbf{r} \times \mathbf{p}$. For simplicity, we consider the single-mode LG beam: $u(r, \phi, z) \propto \exp[i\ell\phi]$. The angular momentum carried by the beam (propagating along the z direction) is then given by

$$j_z = r p_\phi = \varepsilon_0 \omega \ell \, |u|^2 - \frac{1}{2} \varepsilon_0 \omega \sigma_z r \frac{\partial |u|^2}{\partial r}. \tag{3.9}$$

Therefore the total angular momentum is represented by a sum of SAM (2nd term) and OAM (1st term). Namely, we can associate the topological charge ℓ to the OAM. To get the relationship more directly, we introduce the ratio between the angular momentum and energy within the beam cross-section as

$$\frac{J_z}{W} = \frac{\int j_z r dr d\phi}{\int w_E r dr d\phi} = \frac{\ell \int |u|^2 \, r dr d\phi}{\omega \int |u|^2 \, r dr d\phi} - \frac{\sigma_z \int \frac{\partial |u|^2}{\partial r} r^2 dr d\phi}{2\omega \int |u|^2 \, r dr d\phi}$$
$$= \frac{\ell}{\omega} + \frac{\sigma_z}{\omega} \longrightarrow \frac{\hbar \ell}{\hbar \omega} + \frac{\hbar \sigma_z}{\hbar \omega}. \tag{3.10}$$

The LG beam thus carries total angular momentum consisting of OAM ($\hbar\ell$) and SAM ($\hbar\sigma_z$) per photon.

3.1.4 Polarization Vortex

In this subsection, we consider the topological beams that possess a spatial variation of the polarization with a singularity, which is called polarization vortices or vector beams. Especially, we focus on the so-called axisymmetrically-polarized beams,

Table 3.2 ParTypical examples of Jones vector

Linear polarization (x)	Linear polarization (θ)	Right circular polarization	Left circular polarization
$\begin{bmatrix} 1 \\ 0 \end{bmatrix}$	$\begin{bmatrix} \cos\theta \\ \sin\theta \end{bmatrix}$	$\frac{1}{\sqrt{2}} \begin{bmatrix} 1 \\ i \end{bmatrix}$	$\frac{1}{\sqrt{2}} \begin{bmatrix} 1 \\ -i \end{bmatrix}$

typical examples of which are shown in Fig. 3.1b. The singularities are formed by the orthogonally polarized components associated with the axisymmetrical vector field. As a result, axisymmetrically-polarized beams also show a ring-shaped (doughnut-like) intensity profile similar to the LG beams. We also note that, as mentioned in the previous subsection, the polarization of the beam is associated with SAM. In this sense, the polarization vortex can be called optical spin vortex, which is in good analogy to the spin vortex widely observed in various condensed matter systems.

We first provide several basic descriptions for the polarization. Any polarization in the transverse field of the light beam can be decomposed into two orthogonally polarized components, which can be described by $n = \alpha\mathbf{e_x} + \beta\mathbf{e_y}$, where α and β are complex values and satisfy $|\alpha|^2 + |\beta|^2 = 1$. One of the widely-used formalisms is called Jones matrix, which is denoted by

$$J = \begin{bmatrix} \alpha \\ \beta \end{bmatrix} \longrightarrow n = [\mathbf{e_x}, \mathbf{e_y}]J. \tag{3.11}$$

Typical examples are summarized in Table 3.2.

For the polarization vortices, we should take into account the position dependence of the field. The polarization is then described by $n(x, y)$ or $n(r, \theta)$. In Fig. 3.1b, we show the radially-polarized (RP) and azimuthally-polarized (AP) vortex beams, Jones vectors of which are respectively denoted by $\mathbf{RP} = \begin{bmatrix} \cos\phi \\ \sin\phi \end{bmatrix}$ and $\mathbf{AP} = \begin{bmatrix} -\sin\phi \\ \cos\phi \end{bmatrix}$ in the cylindrical coordinate. It is important to note that these vectors can be decomposed into the optical vortex beams with topological charges $\ell = \pm 1$. We can confirm these relationships as

$$\mathbf{RP} = \begin{bmatrix} \cos\phi \\ \sin\phi \end{bmatrix} = \begin{bmatrix} \frac{e^{i\phi} + e^{-i\phi}}{2} \\ \frac{e^{i\phi} - e^{-i\phi}}{2i} \end{bmatrix} = \frac{1}{2}e^{i\phi}\begin{bmatrix} 1 \\ -i \end{bmatrix} + \frac{1}{2}e^{-i\phi}\begin{bmatrix} 1 \\ i \end{bmatrix},$$

$$\mathbf{AP} = \begin{bmatrix} -\sin\phi \\ \cos\phi \end{bmatrix} = \begin{bmatrix} -\frac{e^{i\phi} - e^{-i\phi}}{2i} \\ \frac{e^{i\phi} + e^{-i\phi}}{2} \end{bmatrix} = i\left(\frac{1}{2}e^{i\phi}\begin{bmatrix} 1 \\ -i \end{bmatrix} - \frac{1}{2}e^{-i\phi}\begin{bmatrix} 1 \\ i \end{bmatrix} \right),$$

showing the combinations of the uniformly distributed left and right circular polarizations. In other words, axisymmetrically-polarized beams can be transferred from the LG beams and vice versa [45].

3.2 Generation of the Topological Light Beams

3.2.1 Generation of LG Beams

This section will introduce several techniques for generating the single-mode LG beams, which are useful and needed in the experimental sections. Since the LG mode is one of the cavity modes of the laser, we can obtain the LG beams directly from the laser. In fact, the vortex beams having such as high power, ultrashort pulse, and specific oscillation wavelengths have been demonstrated from the laser oscillator [26, 27, 52]. It is also noted that the formation of the singularities associated with multiple LG modes of the laser and their nonlinear interactions with the gain media have been attractive research fields in the laser physics and technology [9, 20, 30]. However, for laser spectroscopy, both the controllability and stability (especially for higher modes) of the LG beams are the most important aspects, which are missing in the vortex laser in general.

Alternatively, one can use the mode conversion from the fundamental mode (TEM$_{00}$) of the laser. The most simple way to realize such a mode conversion is to use the optics with spiral phase modulations such as spiral mirror, phase plate, and spatial light modulator [5, 49]. The remarkable advantage of this method is the low optical loss in the conversion, which is useful for the laser spectroscopy that needs a critical power. However, the conversion loss and resultant contamination are usually high in this method because the intrinsic and extrinsic discontinuities cannot be avoided in such spiral modulations with a singularity. The contamination of other modes can be separated by the phase modulation with holographic grating instead of the simple spiral phase modulation, which will be demonstrated in Sect. 3.2.3 [11, 21]. We also review the mode conversion using the Gouy phase shift of the propagating beam in Sect. 3.2.2, which can realize high conversion efficiency in principle [4]. However, since the controllability is unfortunately low in general, this method is not useful for generating the LG beam, but is more effective for the detection. Moreover, the conversion process includes an essential feature of topology in the optical field.

3.2.2 Mode Conversion Using Gouy Phase

As shown in Sect. 3.1.2, HG and LG modes are individual solutions of a paraxial wave equation, each of which forms a complete set of the orthogonal function. Therefore, these two modes can be transferred from each other. Namely, any LG mode can be expressed in terms of HG modes and vice versa. A typical example is shown in Fig. 3.3a), where LG_1^0 is decomposed into a superposition of HG_1^0 and HG_0^1 modes. The imaginary unit on HG_1^0 corresponds to the phase delay of $\exp[i\,(\pi/2)]$ from HG_0^1 mode. Therefore we can obtain the LG_1^0 beam from a HG beam that is rotated by $45°$(i.e. $HG(45°) = HG_1^0(x) + HG_0^1(y)$) with a phase shift of $\pi/2$ between x and y directions. In order to produce such a phase shift, we can employ the Gouy

Fig. 3.3 **a** A typical relationship between a LG mode ($\ell = 1$) and a superposition of HG modes, where the multiplication by i indicates a phase delay of $\pi/2$. **b** An astigmatic mode converter by a cylindrical lens

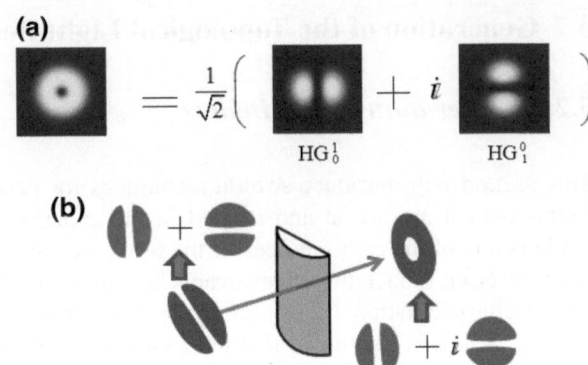

phase $\Phi_G(z)$ of the focused HG beam, which is denoted by the exponential term including $\tan^{-1}(z/z_R)$ in (3.3) and is given by

$$\Phi_G^{HG}(z) = \exp\left[-i\,(1+m+n)\tan^{-1}\left(\frac{z}{z_R}\right)\right]. \tag{3.12}$$

For completion, the Gouy phase of the focused LG beam is given by

$$\Phi_G^{LG}(z) = \exp\left[-i\,(1+2p+|\ell|)\tan^{-1}\left(\frac{z}{z_R}\right)\right]. \tag{3.13}$$

The Gouy phase shift is thus determined by the mode indices and the location of the beam after the lens. For the anisotropic phase shift in the xy plane, the focusing by a cylindrical lens can be used, which is called astigmatic mode converter.

Here we consider the mode conversion from a HG beam with mode indices $(m, n) = (1, 0)$ or $(0, 1)$ to a LG with $\ell = 1$ as shown in Fig. 3.3b. When the HG_0^1 (45°) beam is focused in the x direction by a cylindrical lens, $HG_1^0(x)$ component at the focal point has a phase shift of $\Delta\Phi_G^x(0) = \Phi_G^x(0) - \Phi_G^x(-\infty) = \frac{\pi}{2}$ while $HG_0^1(y)$ has $\Delta\Phi_G^y(0) = \Phi_G^y(0) - \Phi_G^y(-\infty) = 0$, yielding a phase difference of

$$\Delta\Phi_G(0) = \Delta\Phi_G^x(0) - \Delta\Phi_G^y(0) = \frac{\pi}{2}. \tag{3.14}$$

We can thus obtain the LG_1^0 mode at the focal point (but this is not useful for the experimental use and can be improved by introducing a composite lens system [4]). On the other hand, when rotating the cylindrical lens by 90°, the HG_0^1 (45°) beam is focused in the y direction and shows a phase difference of $\Delta\Phi_G(0) = -\frac{\pi}{2}$, leading to the LG_{-1}^0.

For the actual experimental setup, a pair of cylindrical lenses are introduced to generate the LG mode outside of the Rayleigh region [4]. In this case, the separation of the cylindrical lenses is optimized such that the two orthogonal HG components satisfy $\Delta \Phi_G(\infty) = \dfrac{\pi}{2}$. As a general definition, the HG_n^m modes are converted into LG_ℓ^p modes with transformations $\ell = m - n$ and $p = \min(m, n)$. We also note that in this method, high-order HG beams should be prepared before the conversion for generating arbitrary LG beams. However, high-order HG beams can be obtained by using various methods that is much easier than generating higher LG beams [39, 53].

3.2.3 Mode Conversion Using Holographic Grating

The most widely used mode conversion technique is to use the holographic grating recorded in an optical element such as a phase plate or liquid crystals. For the LG beam generation, the interference pattern between a LG mode and an inclined plane-wave is commonly used, where the LG beam acts as an object beam and the plane-wave acts as a reference beam for holography [11, 21]. Since the reconstruction of the object field encoded within the hologram is based on the diffraction, the LG beam can be obtained as a distinct mode. In other words, we can spatially separate the contamination from the LG beam of interest, which is a significant advantage of this method.

The controllability of the mode conversion using the holographic gratings can be realized by a computer-generated hologram (CGH) displayed on a spatial light modulator (SLM). Since the SLM is usually composed of liquid crystals, we have to take care of its low damage threshold for the laser and slightly high optical loss in the presence of the electrodes. We also take care of the diffraction efficiency which reduces the overall efficiency in the conversion. Despite of these disadvantages, the use of SLM that allows to generate the variable LG beams are highly useful for the laser spectroscopy. The dynamic control of the beam and the generation of specific beams such as multiple LG beams can also be achieved by the SLM.

Here we describe the formulations of the CGH for a simple example; the conversion from the fundamental Gaussian beam to a single-mode LG beam. The electric fields of the LG beam (E_{LG}) and the plane-wave beam (E_G) with a small tilting angle θ are described by

$$E_{LG}(r, \phi, z) = u_{LG} \exp\left[i\ell\phi + ikz\right], \quad E_G(x, y, z) = u_G \exp\left[i(kz - k\theta x)\right],$$
(3.15)

respectively. The interference intensity $I(r, \phi, z_0)$ is thus given by

$$I(r, \phi, z_0) = |E_{LG}|^2 + |E_G|^2 + E_{LG}E_G^* + E_{LG}^* E_G$$
$$= I_{LG} + I_G + 2u_{LG}u_G \cos(\ell\phi + k\theta x).$$
(3.16)

This forms fork-shaped interference fringes with a spacing of $2\pi/k\theta$ in the x axis (except for the singularity) as shown in Fig. 3.4. The electric field of the

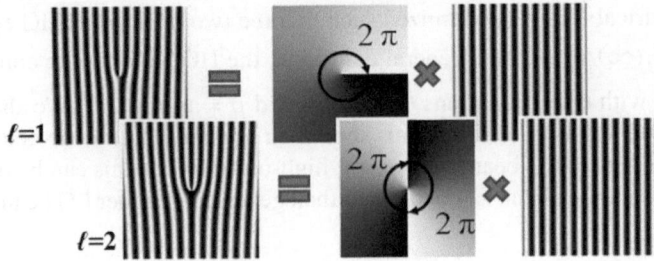

Fig. 3.4 Fork-shaped holographic gratings

transmitted Gaussian beam (TEM$_{00}$) with the same angle θ through the hologram is then described by

$$E_t \propto E_{\mathrm{G}} \mathrm{I} = (\mathrm{I}_{\mathrm{LG}} + \mathrm{I}_{\mathrm{G}}) E_{\mathrm{G}} + E_{\mathrm{LG}} \mathrm{I}_{\mathrm{G}} + E_{\mathrm{LG}}^* E_{\mathrm{G}} E_{\mathrm{G}}. \qquad (3.17)$$

In the equation, the first term corresponds to the 0th order diffraction, and the other two terms are the ±1st order of diffraction, respectively, where one can see that the second term corresponds to the LG beam with a topological charge ℓ.

To increase the purity of the output beam, the nonlinearity of the gradation in the CGH should be taken into account for the modulation. The conversion efficiency can be enhanced by optimizing the modulation pattern. Moreover, the diffraction characteristics of the holographic grating allow to generate the LG beam with different ℓ in the different directions, which can be applied to the multiple information recording, transfer and processing [3, 7, 13, 18, 31].

3.2.4 Generation of Polarization Vortex Beam

Similar to the optical vortices (LG beams), the generation of the polarization vortices can be realized either inside or outside of the laser oscillator. Here we introduce an external conversion technique for generating the axisymmetrically-polarized beams by using spatially-variant optical elements such as axisymmetric polarizer (ASP) and axisymmetric waveplate (AWP) [42, 55]. These elements are made of azimuthally periodic array of the elements and are commercially available. The Jones matrices of ASP and AWP are yielded by

$$\mathrm{ASP} = \begin{bmatrix} \cos^2 \phi & \frac{1}{2} \sin 2\phi \\ \frac{1}{2} \sin 2\phi & \sin^2 \phi \end{bmatrix}, \quad \mathrm{AWP} = \begin{bmatrix} \cos \phi & \sin \phi \\ \sin \phi & -\cos \phi \end{bmatrix}, \qquad (3.18)$$

where ϕ is the azimuthal angle in a global coordinate, indicating the spatially-variant polarizations. When passing the uniformly right-circular polarized beam through the ASP, we obtain

$$\frac{1}{\sqrt{2}} \begin{bmatrix} \cos^2 \phi & \frac{1}{2}\sin 2\phi \\ \frac{1}{2}\sin 2\phi & \sin^2 \phi \end{bmatrix} \begin{bmatrix} 1 \\ i \end{bmatrix} = \frac{1}{\sqrt{2}} e^{i\phi} \begin{bmatrix} \cos \phi \\ \sin \phi \end{bmatrix}, \tag{3.19}$$

which corresponds to the RP beam. Similar conversion can be achieved when passing the linearly polarized beam through the AWP. We note that in (3.19), the resultant RP beam exhibits a rotational phase, showing a characteristic of the optical vortex. On the other hand, the RP beam by the AWP has no spatial phase variation. The rotational phase in the former case allows one to obtain the optical vortex beam (LG beam) with uniform polarizations by using a polarizer [45].

The polarization angle in the local coordinate can be controlled by a pair of half-wave plates (HWPs) with a relative angle of α. The Jones matrix of a paired HWP is described by

$$R(-\alpha) \cdot HWP \cdot R(\alpha) \cdot HWP = \begin{bmatrix} \cos(2\alpha) & -\sin(2\alpha) \\ \sin(2\alpha) & \cos(2\alpha) \end{bmatrix}, \tag{3.20}$$

using $HWP = \begin{bmatrix} 1 & 0 \\ 0 & \exp(-i\pi) \end{bmatrix}$ and rotation matrices $R(\pm\alpha)$. The transmitting RP beam in (3.19) through the paired HWP is then given by

$$R(-\alpha) \cdot HWP \cdot R(\alpha) \cdot HWP \cdot ASP \cdot \frac{1}{\sqrt{2}} \begin{bmatrix} 1 \\ i \end{bmatrix} = \frac{1}{\sqrt{2}} e^{i\phi} \begin{bmatrix} \cos(2\alpha + \phi) \\ \sin(2\alpha + \phi) \end{bmatrix}. \tag{3.21}$$

This Jones vector represents an axisymmetrically-polarized beam, the polarization of which is determined by both the global and the local azimuthal angles. For example, we can obtain the RP beam at $\alpha = 0$, whereas the AP beam is generated at $\alpha = \pi/4(45°)$. An example of the optical setup is shown in Fig. 3.5, together with typical output polarization distributions. We can also remove the rotational phase

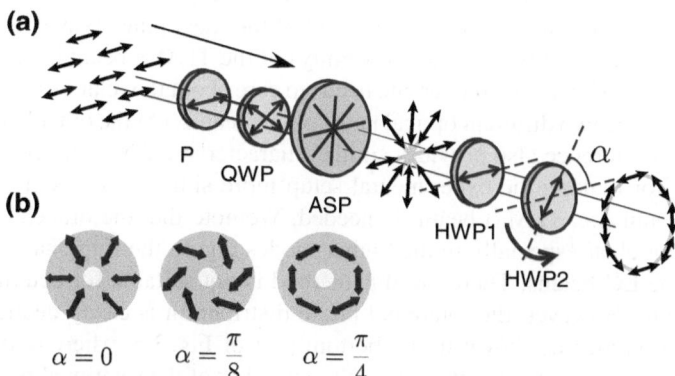

Fig. 3.5 a Optical setup for generating polarization vortex from a uniformly linear polarized beam (*P* polarizer, *QWP* quarter-wave plate, *ASP* axisymmetric polarizer, *HWPs* half-wave plates). **b** Examples of the polarization distribution of the polarization vortex generated from **a**

variation by using the AWP with linearly polarized beam. In this case, we obtain the axisymmetrically-polarized beam given by

$$R(-\alpha) \cdot HWP \cdot R(\alpha) \cdot HWP \cdot AWP \cdot \begin{bmatrix} 1 \\ 0 \end{bmatrix} = \begin{bmatrix} \cos(2\alpha + \phi) \\ \sin(2\alpha + \phi) \end{bmatrix}. \qquad (3.22)$$

3.3 Evaluation of the Topological Light Fields

To analyze the topological light fields including not only the light sources, but also the signal fields in the laser spectroscopy, it is necessary to employ the evaluation techniques. Since the topological beams are characterized by their spatial properties such as phase and polarization distributions, several kinds of spatially-resolved technique have been proposed. In this section, we review the two major techniques that are useful for the laser spectroscopy using optical vortex pulses. The most widely used technique for evaluating the optical vortex fields is based on the interference. Because of the singularity of the optical vortex, the interference patterns are significant so as to qualitatively characterize the vortex fields (Sect. 3.3.1). Another major technique is based on mode conversions, which allow a quantitative analysis using simple experimental setup (Sect. 3.3.2).

3.3.1 Interference Analysis

The interference between a single-mode LG beam and a tilted plane-wave beam produces a characteristic fringe pattern (fork-shaped interference fringes) as discussed in Sect. 3.2.3. Therefore we can easily determine the topological charge ℓ from the branch of the fork-shaped fringes. In the practical experiment, a Gaussian beam (TEM$_{00}$) is used as a reference beam instead of the ideal plane-wave beam. In this case, the quality (beam profile and uniformity) of the TEM$_{00}$ beam is important for evaluating the LG beam. Moreover one needs to care about the relative phase fluctuations between the two different optical paths of the beams. On the other hand, the use of the self-interference also provides similar characteristic fringe patterns (Fig. 3.6) and allows for making the experimental setup more simple because no additional reference beam (i.e. TEM$_{00}$ beam) is needed. We note that the branch of the fork pattern is not clear especially in the higher modes due to the ring-shaped intensity profile of the LG beams. Therefore this method is not suitable for determining the singular point. However, the rotational phase distribution is easily deduced by the coaxial interference as shown in the bottom part of Fig. 3.6, where only the rotational fringes remain in the pattern. From the number of the rotational fringes n, the topological charge ℓ is determined by a simple relationship of $n = 2|\ell|$.

Next we consider the case of evaluating the multi-mode LG beams (multiple or composite optical vortices). In this case, the interference pattern becomes

Fig. 3.6 Cross-sectional intensity distributions of various LG beams (*top*) and their self-interferences by mean of non-coaxial (*middle*) and coaxial (*bottom*) configurations

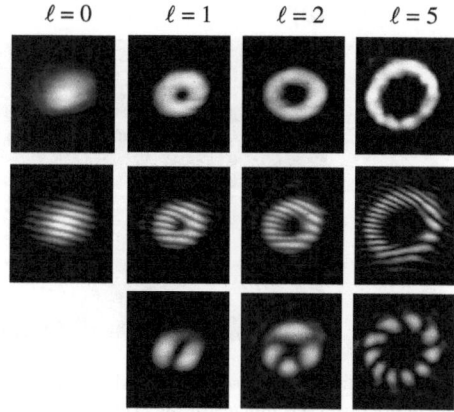

$\ell = 0 \qquad \ell = 1 \qquad \ell = 2 \qquad \ell = 5$

complicated even in the interference with a well defined Gaussian beam, and we cannot deduce the phase distribution directly from the interference. Instead, Fourier analysis of the image can be used to select (reconstruct) the phase distribution [25]. We consider the multi-mode LG beam, which is expanded by a series of LG modes as follows:

$$E_{\mathrm{obj}}(r, \phi, z, t) = \sum_{p=0}^{\infty} \sum_{\ell=-\infty}^{\infty} a_{\ell,p} u_{\ell,p}(r, z) \exp\left[i\ell\phi + i\left(kz - \omega t\right)\right], \quad (3.23)$$

where $u_{\ell,p}(r, z) \exp\left[i\ell\phi\right]$ corresponds to (3.4). Here the radial index p is take in to account to complete the analysis. When the multi-mode LG beam is interfered with a tilted TEM$_{00}$ beam ($E_{\mathrm{ref}} = b \exp\left[i\left(k\cos\theta z - k\sin\theta x - \alpha\right)\right]$), the interference intensity is given by

$$I(r, \phi, z_0) = |b|^2 + \sum_{p=0}^{\infty} \sum_{\ell=-\infty}^{\infty} \left|a_{\ell,p} u_{\ell,p}(r, z_0)\right|^2$$

$$+ b^* \sum_{p=0}^{\infty} \sum_{\ell=-\infty}^{\infty} a_{\ell,p} u_{\ell,p}(r, z_0) \exp\left[i\left(\ell\phi + k\sin\theta x + \beta\right)\right]$$

$$+ c.c. \qquad (3.24)$$

The Fourier transform of (3.24) produces 3 distinct peaks of $I_{\mathrm{DC}}^F(k_x \approx 0, k_y \approx 0)$, $I_{\mathrm{AC+}}^F(+k\sin\theta, 0)$, and $I_{\mathrm{AC+}}^F(-k\sin\theta, 0)$ in the spatial frequencies space (k_x, k_y). When we select the $I_{\mathrm{AC+}}^F(+k\sin\theta, 0)$ component and perform its inverse Fourier transform, we obtain

(a)   **(b)**   0   2π

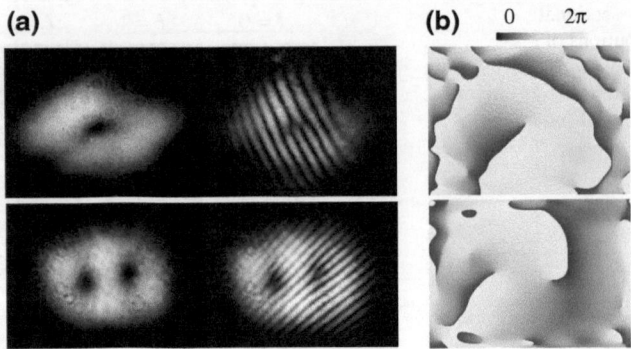

Fig. 3.7 Examples of Fourier analysis of the interference for composite optical vortices. **a** Experimentally observed images of the composite optical vortices (*left*) and their interferences with the TEM$_{00}$ beam (*right*). **b** Phase distributions reconstructed from the interferences

$$I_{AC+}(r, \phi) \exp[-ik \sin \theta x] \propto \sum_{p=0}^{\infty} \sum_{\ell=-\infty}^{\infty} a_{\ell,p} u_{\ell,p}(r) \exp[i(\ell\phi)]$$

$$= \sum_{p=0}^{\infty} A_{\ell,p}(r, \phi). \tag{3.25}$$

The imaginary part of $A(r, \phi)$ provides the spatial phase distribution of the initial multi-mode LG beam. In (3.25), $\exp[-ik \sin \theta x]$ term shifts the spatial frequency of I_{AC+} to the DC position ($k_x \approx 0$, $k_y \approx 0$), indicating that the interference component arising from a tilt of θ in the reference TEM$_{00}$ beam is removed. Here we show the results of the phase distributions reconstructed from the experimentally observed interference patterns in Fig. 3.7, where the optical vortices with $\ell = 1$ (upper) and $\ell = 2$ (lower) are observed. We can verify that the fork-like fringe patterns in (a) result in a singular point(s) of the rotational phase in (b). It is important to note that the results of the $\ell = 2$ case show a split of the phase singularity, which is associated with the weak perturbation of the $\ell = 0$ component. Such perturbation has often been observed in laser spectroscopy and plays an important role to characterize the interactions with materials. The reconstructed phase distribution allows us to address the position of singularities very precisely even in the multi-mode LG beam.

Finally, we comment on the fact that (3.25) corresponds to a Fourier series of the LG mode functions in terms of the topological charge ℓ. The relative amplitude of the mode is thus obtained by

$$a_{\ell,p} = \frac{1}{2\pi} \int_0^{\infty} \int_0^{2\pi} A_p(r, \phi) \exp[i(\ell\phi)] d\phi r dr. \tag{3.26}$$

If we neglect the radial mode (i.e. $p = 0$), (3.26) represents the OAM resolved spectrum of the electric field, where the power spectrum in terms of the LG$_{\ell}$ modes can be evaluated by $|a_\ell|^2$. It is important to note that $a_{\ell,p}$ provides both the amplitude and phase of the topological charge ℓ. Therefore we can also evaluate the relative phase between the modes, which is a unique feature of this technique [25].

3.3.2 OAM Resolved Spectroscopy Based on Mode Conversion

The mode indices of the single-mode HG beam are easily deduced from the beam profile, which is much easier than the case of the single-mode LG beam. In Fig. 3.2, the beam profile of the LG beam is always characterized by a central dark hole, which is not responsible for the topological charge dependence. On the other hand, the numbers of dark stripes along x and y in the HG_m^n correspond to its mode indices n, and m, respectively. Such a significant relationship in the HG beam provides an idea of using the mode conversion technique to determine the topological charge of the LG beam, which is given by a reverse conversion process in Sect. 3.2.1. The most simple way is to use a cylindrical lens to convert the LG beam into an astigmatic beam with dark stripes. The number of the stripes is equal to the topological charge ℓ [16].

The difference of the beam profile is also remarkable between the lowest-order TEM_{00} mode and higher-order LG modes. Among a series of the LG modes, only TEM_{00} has a single peak at the beam center. Therefore we can select the TEM_{00} mode from multiple or composite optical vortices by using a spatial (pinhole) filter or by propagating through a single mode fiber. The most widely-used technique for analyzing optical vortex fields is based on this procedure [31]. The technique measures the intensity of the TEM_{00} component that is obtained by the mode conversion of the LG beams using an approach that is the reverse of that used for generating an LG beam. By changing the mode of interest for conversion, we can obtain an OAM-resolved (ℓ-resolved) spectrum, in which the analyzed beam is decomposed into a series of LG modes with corresponding intensities. The converted fundamental TEM_{00} beam is also useful in terms of adopting the technique for various applications, such as telecommunication, optical storage, and laser spectroscopy [3, 8, 46, 47, 50, 51].

Similar to LG beam generation from laser (Sect. 3.2.3), $LG_\ell \rightarrow TEM_{00}$ conversion is realized by a CGH on SLM. From (3.17), the 1st-order diffraction of the incident field E_i is given by $E_t^{+1} \propto E_{LG} E_G^* E_i = E_\ell E_0^* E_i$, where E_ℓ and E_0 corresponds to the LG and plane-wave fields of the CGH. We assume that E_i is a multi-mode optical vortex with topological charges $\sum m$, and is expressed by

$$E_i = \sum_{m=-\infty}^{\infty} a_m u_m(r, z) \exp\left[im\phi + i(kz - k\theta x)\right]. \qquad (3.27)$$

The electric field of the diffraction beam through the CGH is then described by

$$E_t^{+1} \propto u_\ell u_0 \sum_{m=-\infty}^{\infty} a_m u_m(r, z) \exp\left[i\left\{(\ell + m)\phi + kz + \alpha\right\}\right], \qquad (3.28)$$

indicating that each of m-components is converted to the field with $\ell + m$. We can thus select the component that satisfies $\ell + m = 0$ (i.e. TEM_{00} beam) using a spatial filtering method mentioned above, and obtain the OAM resolved spectrum by changing ℓ in CGH. We note that the OAM spectrum obtained by this procedure is a power spectrum of $|a_m|^2$, where the phase information of a_m is lost. In contrast, the Fourier analysis of the interference in Sect. 3.3.1 provides complete sets of a_m.

As a final note in this section, we comment on the precision of the OAM spectrum, which highly depends on the conversion efficiency and its uniformity for various OAM [2, 14, 44]. For example, the spatial filtering of the TEM_{00} component is based on the difference between the spatial intensity profiles of TEM_{00} and LG beams. Therefore, the imperfect components of the converted TEM_{00} reduce the accuracy of the spectrum. Furthermore, owing to the diffractive optical element in the techniques, the degree of conversion usually shows a strong OAM dependence with a fixed propagation distance, leading to variation of the conversion efficiency. These problems can be improved by using an spiral phase modulation that allows both focusing and mode conversion simultaneously in simpler optical setup [44]. Moreover, by taking advantage of the flexibility of the SLM, the focusing length can be controlled to compensate the conversion efficiency for various OAM.

3.4 Application to the Four-Wave Mixing (FWM) Spectroscopy

In this section, we demonstrate the coherent OAM transfer from the optical vortex pulses to the center of mass momentum of excitons in semiconductor GaN. We show that the four-wave mixing (FWM) signal from GaN excitons is spatially coherent within the exciton dephasing time and exhibits the OAM (phase singularities) satisfying the OAM conservation law. We also demonstrate the time evolution of the exciton OAM, and the optical gate control by changing the topological charge of the LG beams.

3.4.1 FWM Spectroscopy of Excitons Using Ultrashort Pulses

We briefly overview the degenerate FWM (DFWM) technique for semiconductor excitons [40]. For convenience, we sometimes call DFWM just FWM hereafter. In the technique, a pair of laser pulses (k_1 and k_2) with a time separation of τ is introduced onto a sample, where τ is defined to be positive when k_1 pulse precedes the k_2 pulse. When the laser is resonant with the transition from the exciton vacuum state ($|G\rangle$) to the first excited exciton state ($|X\rangle$), the first pulse (k_1) creates the exciton

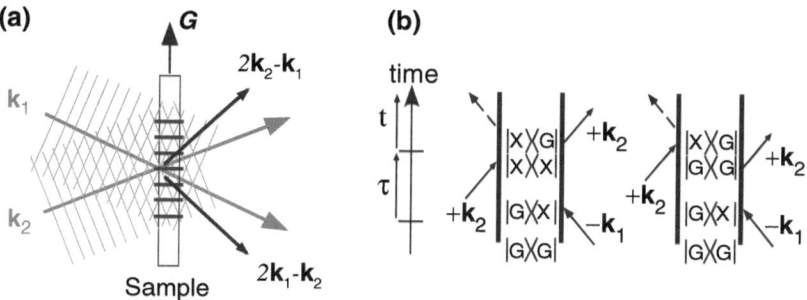

Fig. 3.8 a Schematic illustration of degenerate FWM, where G represents a polarization grating induced by $\mathbf{k}_1$ and $\mathbf{k}_2$ pulses. **b** Feynnman diagrams for a 2-level system in FWM

polarization, which undergoes dephasing after the excitation. If the second pulse ($\mathbf{k}_2$) arrives during the exciton dephasing time, the interference of the polarizations created by $\mathbf{k}_1$ and $\mathbf{k}_2$ pulses result in a polarization grating (Fig. 3.8a). The self-diffraction signal of the second pulse is generated in $2\mathbf{k}_2 - \mathbf{k}_1$ direction due to third-order nonlinearity of the exciton transitions, which is integrated over the real time for detection (TI-FWM). The reverse process ($\mathbf{k}_2$ pulse precedes the $\mathbf{k}_1$ pulse) also generates the signal in $2\mathbf{k}_1 - \mathbf{k}_2$ direction. We can thus spatially separate the FWM signal from the excitation beams even in the resonant condition, which is a remarkable advantage of the DFWM spectroscopy.

For semiconductor excitons, the manifestation of the optical nonlinearity usually arises from the phase-space filling and Coulomb interactions. The dephasing processes of exciton including dephasing time T_2 are evaluated by the time evolution of the TI-FWM signal, which can be obtained by changing τ. The TI-FWM signal intensity is thus given by

$$I_{\mathrm{FWM}}(\tau) \propto \int |\mathbf{P}^{(3)}(\tau, t)|^2 dt. \tag{3.29}$$

Here $\mathbf{P}^{(3)}(\tau, t)$ is the third-order nonlinear polarization, which is given by $\mathbf{P}^{(3)} = \langle \mu_{\mathrm{XG}} \rho^{(3)}(t) \rangle$, where μ_{XG} is the exciton dipole moment and $\rho^{(3)}(t)$ is the third-order density matrix which is determined by the semiconductor optical Bloch equation. Both the phase-matching condition of the third-order polarization and the rotational-wave approximation allow two types of the polarizations to survive, which can be summarized by the Feynman diagrams in Fig. 3.8b.

We now describe the formulation of $\mathbf{P}^{(3)}(\tau, t)$. Here we assume the homogeneous broadening of the exciton state. Within the framework of the optical Bloch equation using the δ-function pulses with a collinear polarizations $E_1(\mathbf{k}_1, \omega)$ and $E_2(\mathbf{k}_2, \omega)$ for excitation, the third-order polarization $P^{(3)}_{2k_2 - k_1}$ is described by

$$P^{(3)}_{2k_2 - k_1}(t \geq \tau, \tau) = \frac{i}{\hbar^3} E_2^2 E_1^* \exp\left[i\left(2\mathbf{k}_2 - \mathbf{k}_1\right) \cdot \mathbf{r} - i\omega t\right] \alpha(t, \tau), \tag{3.30}$$

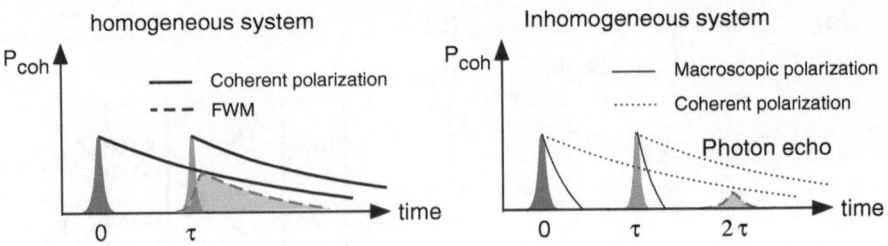

Fig. 3.9 Schematic illustrations of the time-resolved FWM as a function of time t. Note that τ indicates the delay time of the probe pulse relative to the first pump pulse. In TI-FWM, the signal is obtained from the integration of the FWM over $t \geq \tau$. To consider the possible processes, the FWM in the inhomogeneous system is also shown, where the FWM appears as a photon echo, resulting in a difference in the definition of T_2

where $\alpha(t, \tau)$ represents the term related to the exciton, and is given by

$$\alpha(t, \tau) = |\mu_{XG}|^4 \exp(+i\omega_X\tau) \exp(-\gamma\tau) \exp(-i\omega_X t) \exp(-\gamma t)$$
$$= |\mu_{XG}|^4 \exp[-i\omega_X(t - \tau)] \exp[-\gamma(t + \tau)]. \quad (3.31)$$

Here ω_X and γ are the exciton resonant energy $\hbar\omega_X$ and dephasing rate, respectively. The FWM signal thus propagates along the direction of $\mathbf{k}_{FWM} = 2\mathbf{k}_2 - \mathbf{k}_1$. The TI-FWM signal intensity is then given by

$$I_{FWM}(\tau) \propto \frac{|\mu_{XG}|^8}{\hbar^6} e^{-2\gamma\tau} \int_0^\infty dt \, |\exp(-i\omega_X t) \exp(-\gamma t)|^2 \propto e^{-2\gamma\tau}. \quad (3.32)$$

We thus observe the time evolution of the signal with a decay rate of 2γ, which can be represented by $I_{FWM} = I_0 \exp(-2\tau/T_2)$ using a dephasing time T_2 (Fig. 3.9). In the case of the single-mode LG beams for excitation, (3.30) is modified to

$$P^{(3)}_{2k_2-k_1}(t, \tau) = \frac{i}{\hbar^3} E_2^2 E_1^* \exp[i\{\mathbf{k}_{FWM} \cdot \mathbf{r} + (2\ell_2 - \ell_1)\phi\} - i\omega t]\alpha(t, \tau). \quad (3.33)$$

The equation indicates that the FWM signal with a topological charge (OAM) of $\ell_{FWM} = 2\ell_2 - \ell_1$ is generated if the exciton term $\alpha(t, \tau)$ maintains the spatial coherence.

The (3.33) clearly indicates that the FWM signal is characterized by the wavevector ($\mathbf{k}_{FWM} = 2\mathbf{k}_2 - \mathbf{k}_1$) and OAM ($\ell_{FWM} = 2\ell_2 - \ell_1$) by using single-mode LG pulses for excitation. Namely, we can generate the optical vortices with an arbitrary OAM in the same direction by changing the OAM in the one of the incident pulses. It is important to note that the OAM of the FWM signal is affected by the spatial coherence of exciton center of mass momentum. In (3.31), μ_{XG} may exhibit spatial variations in terms of spatial coherence, which changes the OAM of the FWM signal from $\ell_{FWM} = 2\ell_2 - \ell_1$. If the spatial dephasing such as diffusion or spatially-dependent scattering is efficient, the OAM of the FWM signal also decays with τ.

3.4.2 FWM Spectroscopy Using Optical Vortex Pulses

Here, we will demonstrate the coherent OAM transfer from optical vortex (LG) pulses to the center of mass momentum of excitons in semiconductor GaN [28, 50]. The samples studied here are a GaN film (thickness: 3 μm) grown on (0001) c-plane sapphire substrate and a free-standing bulk GaN (thickness 70 μm). The GaN exciton has a large oscillator strength, enabling us to obtain the efficient FWM signal because I_{FWM} is proportional to the fourth power of the transition oscillator strength (see (3.32)). The FWM measurement was performed using a frequency-doubled mode-locked Ti:sapphire laser with the spectral width of 10 meV ($\sim$1.0 nm). Without spatial modulations for the LG beams, the pulse duration is estimated to be 130–160 fs. The center energy of $\sim$3.5 eV ($\sim$355 nm) was selected for the GaN exciton transitions. The laser pulse is split into two collinearly polarized pulses with nearly equal intensities. The delay time τ between the two pulses was determined by the variable spatial delay line. The mode conversion technique using CGH on SLM or AWP was used for generating the LG beams. The two pulses were overlapped on the sample using a lens ($f = 200$ mm), and the FWM signal was collected in the reflection geometry. The sample was mounted on a cold finger of a liquid helium cryostat and whole data were taken at $\sim$10 K.

Figure 3.10a shows a typical FWM spectrum taken at $\tau = 0$, where the laser spectrum is also shown as a dashed curve. In the FWM spectrum, the lower- and higher-energy peaks are identified as electron-heavy-hole-exciton transition (A- exciton: X_A) and electron-light-hole-exciton transition (B- exciton: X_B), respectively. The simultaneous excitation of the two transitions results in a coherent oscillation (generally called quantum beat: QB) in the time-evolution of the FWM as shown in Fig. 3.10b, where the beating period (T_{QB}) of 0.51 ps is consistent with the energy separation of 8.3 meV ($\Delta\nu \sim 2$ THz). Assuming a homogeneously broadened exciton system, the FWM signal with a QB oscillation can be expressed by

$$I_{FWM} = I_0 \left[1 + \cos\left(2\pi/T_{QB}\right)\right] \exp(-2\tau/T_2), \tag{3.34}$$

where we neglect the Coulomb interactions between excitons. From the fitting to the data, we obtain the dephasing time of $T_2 = 1.40$ ps ($T_2 = 2.80$ ps for inhomogeneous case). In Fig. 3.10b, we plot the total intensity obtained by the spatial integration of the beam profile as a function of τ both in the signal directions $2\mathbf{k}_2 - \mathbf{k}_1$ and $2\mathbf{k}_1 - \mathbf{k}_2$. Here a combination of the TEM$_{00}$ pulse ($\ell_1 = 0$) in $\mathbf{k}_1$ direction and LG$_1$ pulse ($\ell_2 = 1$) in $\mathbf{k}_2$ direction was used for excitation. The crossing point between two data sets indicates the zero position of the delay τ.

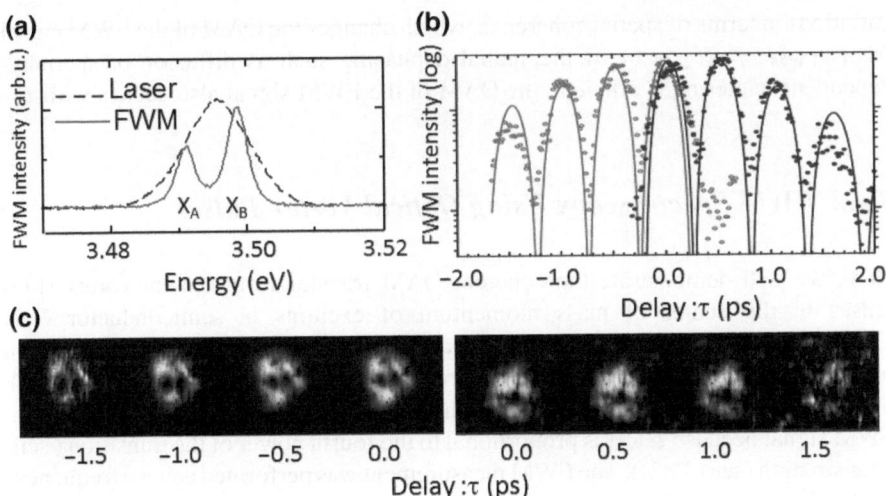

Fig. 3.10 Results of FWM spectroscopy for excitons in a GaN film at 10 K. A combination of TEM$_{00}$ pulse ($\ell_1 = 0$, $\mathbf{k}_1$) and LG (optical vortex) pulse ($\ell_2 = 1$, $\mathbf{k}_2$) was used for excitation. **a** A typical FWM spectrum (*solid line*) at $\tau \approx 0$ together with an excitation pulse spectrum (*dashed line*), where X$_A$ and X$_B$ exciton transitions are simultaneously excited. **b** Plot of FWM signals (logarithm scale) as a function of τ obtained in the direction of $2\mathbf{k}_2 - \mathbf{k}_1$ (signal is dominant at $\tau \geq 0$) and $2\mathbf{k}_1 - \mathbf{k}_2$ ($\tau \leq 0$). The oscillations are attributed to the QBs between X$_A$ and X$_B$ transitions. The *solid lines* show the fits using (3.34). **c** Intensity distributions of FWM signals obtained at each QB oscillation peak in $2\mathbf{k}_2 - \mathbf{k}_1$ (*right*) and $2\mathbf{k}_1 - \mathbf{k}_2$ (*left*), respectively

To investigate the spatial properties of the FWM signal, we first focus on the phase singularity of the beams, which is realized simply by measuring the dark holes in the spatial intensity profile. The FWM signal was recorded using a charge-coupled device (CCD) camera equipped with an image acquisition system. Figure 3.10c shows the spatial profiles of the FWM signal obtained at each QB oscillation peak in the signal directions $2\mathbf{k}_2 - \mathbf{k}_1$ and $2\mathbf{k}_1 - \mathbf{k}_2$, respectively. Each data set shows that the output FWM beam exhibits a characteristic beam profile with dark hole(s) and its intensity decays with increasing $|\tau|$. The presence of the dark hole(s) is responsible for the coherent OAM transfer to the center of mass momentum of excitons in the FWM process. On the other hand, the difference of the beam profiles between the data sets is accounted by the OAM conversion law ($|\ell_{FWM}| = |2\ell_{2(1)} - \ell_{1(2)}| = 2(1)$). The different number of the dark holes (the FWM in $2\mathbf{k}_2 - \mathbf{k}_1$ direction shows a single dark spot at the center while the FWM in $2\mathbf{k}_1 - \mathbf{k}_2$ exhibits two dark holes) can be associated with the different topological charges of the exciton polarization, indicating the OAM conservation during the FWM process. The split of the dark spot in $2\mathbf{k}_1 - \mathbf{k}_2$ is associated with the weak perturbation of the TEM$_{00}$ component ($\ell_{FWM} = 0$), which will be discussed in detail in the next subsection. Such perturbation has widely been observed in the higher charge vortices generated by nonlinear processes.

A more reliable way to check the topological charge of the signal is to observe the interference pattern (Sect. 3.3.1). Figure 3.11b shows the results of intensity profiles

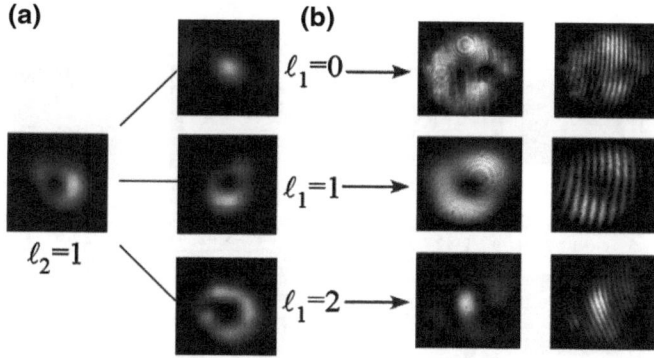

Fig. 3.11 Results of demonstration to show the arbitrary control of the topological charge in the FWM signal by changing the OAM of the excitation pulse. Beam profiles of **a** excitation pulses and **b** corresponding FWM signals at $\tau \approx 0$. The intensity scale varies for each image so as to clarify the dark hole(s). The interference patterns obtained by superimposing the reference TEM_{00} pulse are also shown in the right side of **b**

of the output FWM ($\tau \approx 0$ ps) with $\ell_1 = 0$, 1, 2 together with their interference patterns by overlapping the signal with a tilted TEM_{00} reference pulse. In the measurement, the OAM in $\mathbf{k}_2$ pulse was fixed at $\ell_2 = 1$. Here we also show the intensity profiles of the incident pulses (Fig. 3.11a). On the basis of the conservation law of OAM, the output FWM exhibits $|\ell_{FWM}| = 2$, 1, 0, whose variations are seen in the different number of the dark spots. The phase singularities on the dark spots are confirmed by the fork-typed patterns and agree well with those predicted under the conservation law. It should be noted that these results are also responsible for the arbitrary controls of OAM using FWM process in semiconductor exciton system.

3.4.3 OAM Resolved Spectroscopy of FWM Signal

For a quantitative analysis of the OAM of the generated FWM signal, we further carry out the OAM-resolved (ℓ-resolved) spectroscopy as explained in Sect. 3.3.2, where the FWM signal is decomposed into a series of LG_ℓ modes with corresponding intensities. For the OAM-resolved spectroscopy, the FWM signal in the backward direction was collected and led to the OAM-resolved detection system consisting of a SLM and a spatial filter, and then was detected by a monochromator equipped with a CCD camera. In the results shown below, we focus on the spectrally resolved FWM signal taken at X_A transition.

We first consider the case where the $\mathbf{k}_1$ pulse is a TEM_{00} mode ($\ell_1 = 0$) and the $\mathbf{k}_2$ pulse is a LG mode with $\ell_2 = 1$. Figure 3.12a shows the OAM spectra of the excitation pulses, showing a single peak at the corresponding OAM. The beam profile is also shown in the inset of each spectrum, where a dark singular point at the center

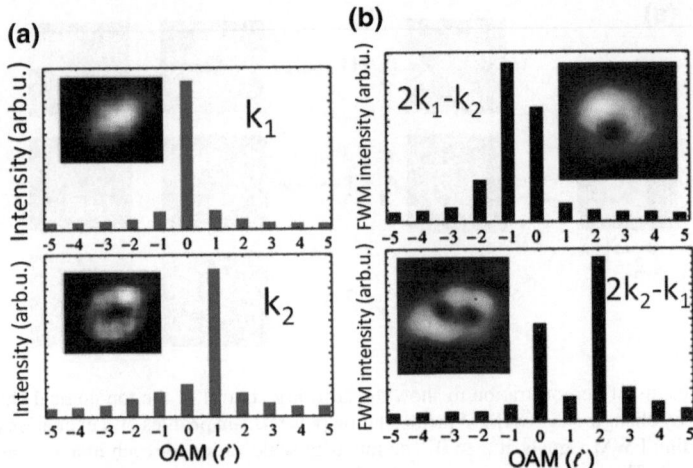

Fig. 3.12 Results of OAM-resolved FWM spectroscopy for GaN excitons (free-standing bulk sample). OAM spectra of **a** excitation pulses and **b** FWM signal of X_A transition at $\tau \approx 0$. Insets show the intensity profiles of the beams

of the beam is clearly observed in the k_2 pulse. However, there still remains weak side modes in the spectrum, which are mainly due to the imperfections of the mode conversion in generation. Moreover, while no side modes are expected in the laser mode, small but finite residual modes are visible in the nearest neighbors of $\ell = 0$, suggesting the imperfect mode conversion in our detection system that determines the background level. It is important to note that the side mode ratio (SMR) highly depends on the conversion system, and we can improve the SMR down to less than 10 dB.

The OAM spectra of the FWM obtained at $\tau \approx 0$ are shown in Fig. 3.12b, together with the CCD images of the signal. On the basis of the OAM conservation law, the ideal conversion provides $\ell_{FWM} = 2 (\equiv 2\ell_2 - \ell_1)$ in $2\mathbf{k}_2 - \mathbf{k}_1$ and $\ell_{FWM} = -1$ $(\equiv 2\ell_1 - \ell_2)$ in $2\mathbf{k}_1 - \mathbf{k}_2$. In each spectrum, an intense peak located at ℓ_{FWM} is observed, whereas another distinct peak is also visible at $\ell_{res} = 0$. The presence of ℓ_{res} component is consistent with the split of $\ell_{FWM} = 2$ into two distinct dark holes with $\ell = 1$ in the signal, which is manifested by the results shown in the insets.

The presence of ℓ_{res} component of the FWM signal can be associated with the $\ell = 0$ components of the incident pulses. Although $\ell = 0$ in the $\mathbf{k}_1$ pulse is small compared with the dominant component of $\ell = -1$ (see Fig. 3.12a), we have to take into account that the peak intensity of the $\ell = 0$ component is located at the center portion of the beam. Because of the third-order nonlinearity of the FWM, the contribution of the $\ell = 0$ component to the signal is much larger than that of the $\ell \neq 0$ components whose peak intensities are spatially distributed along the ring (around the vortex). Moreover, the spatial mode overlap between the individual ℓ_{res} components in the two incident pulses is high (much higher than that

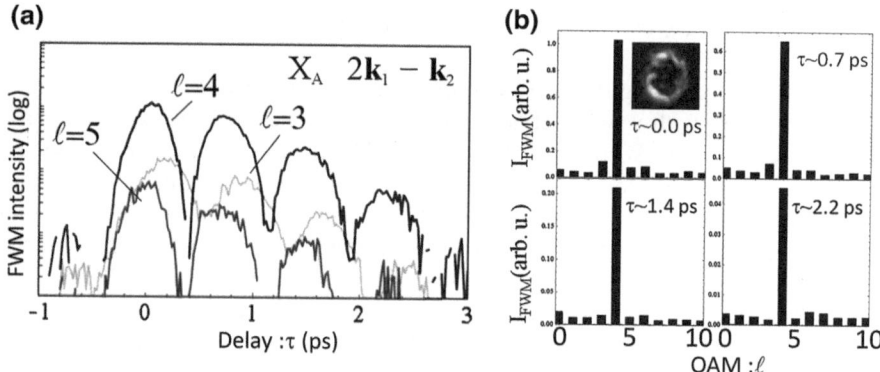

Fig. 3.13 a Time-evolutions of the OAM-resolved FWM signal (logarithm scale). A combination of LG pulses ($\ell_1 = -2$, $\mathbf{k}_1$) and ($\ell_2 = 1$, $\mathbf{k}_2$) was used for excitation. **b** OAM spectra of the FWM signal obtained around the QB oscillation peaks

between ℓ_1 and ℓ_2), increasing the FWM efficiency. For these reasons, the unfavorable $\ell = 0$ component becomes efficient in the FWM conversion process when one of the incident pulses includes $\ell = 0$ component.

The above considerations are verified by another measurement using the incident pulses with $\ell_{1,2} \neq 0$. Figure 3.13a shows the time evolutions of the FWM signal ($2\mathbf{k}_1 - \mathbf{k}_2$) with selected OAM values ($\ell = 3, 4, 5$), where the OAM of the $\mathbf{k}_1$ and $\mathbf{k}_2$ pulses are $\ell_1 = -2$ and $\ell_2 = 1$, respectively. All data show an oscillation of the QB arising from the simultaneous excitation of X_A and X_B transitions mentioned in the previous subsection. Figure 3.13b shows the OAM spectra of the signal ranging from $\ell = 0$ to $\ell = 10$. Each spectrum is obtained at various τ located at the peak of the QB oscillation, and the intensity is normalized by the peak intensity in each spectrum. The value of the vertical axis represents the relative intensity normalized to the peak obtained at the 4th QB peak. The single peak in the spectra indicates a FWM with a well defined OAM that satisfies OAM conservation law, which is also confirmed by the signal intensity profile shown in the inset. The combination of $\ell_1 = -2$ and $\ell_2 = 1$ results in a FWM signal with $\ell_{\text{FWM}} = 4 (\equiv 2\ell_2 - \ell_1)$ only when the excitons keep the OAM transferred by the $\mathbf{k}_1$ pulse and then succeed in converting the OAM via FWM with the $\mathbf{k}_2$ pulse. The results thus indicate both the coherent OAM transfer and near perfect OAM conversion are realized in the exciton states whose spatial coherence remains high during the dephasing time.

It should be noted that the dephasing of OAM observed here is responsible for the spatial dephsing characterized by the azimuthal direction. We would recall that the exciton dipole moment μ_{XG} in (3.31) is expressed in terms of the position of the exciton center of mass momentum, and its magnitude reflects the spatial coherence of exciton (i.e. exciton OAM state) transferred by the $\mathbf{k}_1$ pulse. If the spatial dephasing is efficient, the magnitude of the total μ_{XG} decays with τ. In the cylindrical coordinate, the FWM intensity can therefore be modified as

$$I_{FWM}(\tau, r, \phi) \propto |\mu_{XG}(r, \phi)|^4 \exp\left[-2\gamma(r, \phi)\tau\right], \qquad (3.35)$$

resulting in a broadening of the OAM spectrum. The observed time-evolution of the OAM spectra with a well-defined OAM thus indicates that the center of mass momentum of excitons keep the OAM transferred by the LG pulses over the dephasing time T_2. This is reasonable because the contributions of the spatial dephasing such as diffusion or spatially-dependent scattering are quite small in the picosecond regime. Therefore we conclude that the exciton OAM state is more stable but the dephasing time is limited by T_2 in the present study using FWM. The small contribution of the spatial decoherence can be evaluated by the broad band OAM spectrum of FWM obtained by using the excitation pulse with multiple OAM [28].

3.5 Application to the Pump-Probe Spectroscopy

In the previous section, we demonstrated a global evaluation of the spatial dephasing dynamics of electrons (excitons) in a semiconductor by using optical vortex pulses. A similar approach is applicable in the pump-probe spectroscopy using polarization vortex pulses. Since the change of optical polarization is associated with the symmetry of electron (or spin) system in materials, the excitation with polarization vortex pulses enables us to analyze the dynamics of the global symmetry. In this section, we demonstrate a global evaluation of quasi one-dimensionally (1D) ordered electrons (so-called charge density wave (CDW)) in a ring-shaped crystal using polarization vortex pulses [46]. We first review the CDW in quasi 1D transition metal chalcogenides (MX_3), and then give an overview of the pump-probe spectroscopy for the CDW system. In the analysis, we compare the electron dynamics probed by the azimuthal and radial polarization excitations. We also compare the results with those in the whisker sample, and show that the ring-shaped CDWs exhibit a well-defined azimuthal symmetry associated with the quasi 1D nature.

3.5.1 Topological CDW in MX₃

In this subsection, we begin with a brief introduction of the CDW in quasi 1D conductors, which is schematically illustrated in Fig. 3.14. A CDW is a modulation of the electron density associated with a lattice modulation of same periodicity. Generally, the electronic band gap in solids arises from a Fourier component of the lattice potential. In quasi-1D metal consisting of chains of equally spaced atoms, a modulation of the lattice with wavelength $\lambda = \pi/k_F$ can produce a band gap at $\pm k_F$ by reducing the conduction electron energy below E_F, where E_F and k_F are Fermi energy and wavevector, respectively (left in Fig. 3.14). At low temperatures, the gain in the conduction electron energy overcomes the energy cost of the lattice deformation. As a result, quasi-1D metals undergo a phase transition (metal to insulator

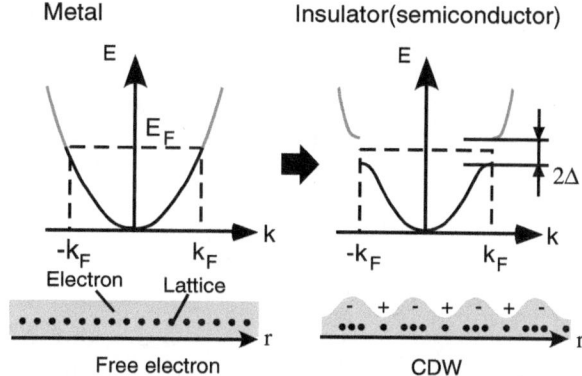

Fig. 3.14 Schematic drawings of a Peierls transition from a normal metallic state to a CDW insulating state in an ideal ID system

transition) when cooled below a certain temperature (T_c). This is a so-called CDW phase transition. Unlike normal insulators (and semiconductors), the ground states of CDW are characterized by two types of collective excitations: phase mode (PM) and amplitude mode (AM). The PM determines the position of the CDW relative to the underlying lattice, and the AM determines the amplitude of the CDW. It should be noted that in realistic quasi-1D metals the three-dimensional (3D) correlation of the 1D CDWs is essential to reduce their fluctuations, meaning that the 3D ordered CDWs exist below T_c.

Among a variety of 1D metals that undergo CDW phase transition, the MX_3 have been most extensively studied over the last few decades [33]. The fascinating feature of MX_3 is provided by a discovery of various types of topological crystals [43]. Because of both the small sizes and damage-free formations of these structures, it is expected to keep the CDW coherence within individual chains with a closed loop, thus making it a good candidate for studying the topological effects on the CDW dynamics [32, 48].

The samples studied here is $NbSe_3$, which is one of the MX_3 family and is the most extensively studied quasi-1D compound. This compound consists of three pairs of metallic chains parallel to the conducting axis (Fig. 3.15a) and undergoes CDW phase transitions at $T_{c1} = 145\,\text{K}$ and $T_{c2} = 59\,\text{K}$. These two transitions are known to take place on two of three types of chains with different nesting conditions in **k**-space. The other chains remain metallic down to the lowest temperatures. Since the CDW with long-range-order requires 3D correlation of electron density between 1D chains, their coherent dynamics attracts considerable interest and has been studied using time-resolved measurements [15, 41].

Figure 3.15b shows several transient reflectivity changes ΔR of a whisker sample for various temperatures below and above T_{c2}. The signal at the lowest temperature is dominated by two features: an exponential decay response (solid line in b) and damped sinusoidal (coherent) oscillations. On the basis of a successful model for the ultrafast optical response of phase transition materials, the exponential decay and oscillation components are attributed to the carrier relaxation across the CDW gap

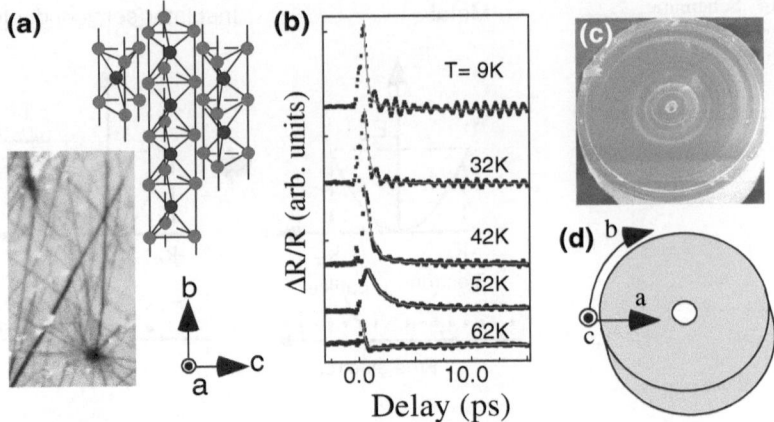

Fig. 3.15 **a** A schematic illustration of NbSe$_3$ whisker and its optical micrograph image. **b** Transient reflectivity changes ΔR of a whisker sample for various temperatures below and above $T_{c2} = 59$ K. The solid lines are the results of exponential fit to the data. **c** Electron micrograph of NbSe$_3$ (disk-like) ring sample, and **d** its schematic illustration of the crystal structure

and coherent CDW motions, respectively [15]. Here, we briefly summarize the carrier relaxation processes. As mentioned above, the 3D correlation of the CDWs forms a gap around E_F below T_c. Therefore one can consider the carrier dynamics in CDW in a similar way to the case of semiconductors, namely, carrier relaxation time (τ_{sp}) reflects the relaxation across the CDW gap via between carriers and carrier-phonon collisions, resulting in a picosecond decay (whereas τ_{sp} is instantaneous on the several tens of femtosecond above T_c, reflecting the intraband relaxation in the metallic phase). Since the CDW gains the gap energy with decreasing the temperature, τ_{sp} shows a characteristic temperature dependence, which can be confirmed in Fig. 3.15b. Another characteristic transient feature is a coherent oscillation resulting from the collective excitations of the CDW. The instantaneously photoexcited carriers change the CDW's equilibrium distributions and result in starting the collective motions. This occurs in the same manner of DE process for coherent phonons. In our previous measurements, a coherent oscillation with $\nu_{AM} \sim 1.1$ THz was observed in NbSe$_3$ and was ascribed to the AM of the CDW [41].

Figure 3.15c shows the scanning electron micrograph of the ring sample used in this study. The chemical vapor transport method was used for sample preparation. Under controlled conditions, thin whiskers naturally form closed-loop crystals by bending and joining the whiskers. Several ring crystals are stacked layer by layer, and form a disk-like structures. It should be noted that T_c's in the ring are similar to those in the whisker. The whisker crystal has a length of a few mm along the conducting b axis (chain axis) while the disk-like ring crystal with a diameter of $\sim$50 μm has a bending b axis along the azimuthal direction in the disk (Fig. 3.15d). The circumference of the center hole of the disk is around 10 μm, which is comparable to the correlation length of CDWs ($\xi_{\parallel b} > 2.5$ μm) .

3.5.2 Pump-Probe Spectroscopy for Quasi 1D CDW System

When we investigate the globally correlated electron systems such as closed-loop CDWs, non-contact and non-destructive measurements are preferable. Therefore, it is advantageous to probe the CDW dynamics with photoexcitation. Ultrafast pump-probe spectroscopy has been widely used in this kind of time-resolved experiments, where optical responses (such as transient reflectivity changes ΔR) of photoexcited electrons are traced by a probe pulse immediately after an intense pump pulse with a delay time between the two pulses.

Generally, the excitation process by the pump can be classified by whether a coherent stimulated-Raman (SRE) excitation or an incoherent dissipative excitation (DE). In the SRE process, various degrees of freedom are coherently excited. The symmetry of the coupling to different degrees of freedom is described via the appropriate Raman tensor. In the DE process, the high energy photoexcited carriers create incoherent excitations by inelastic scattering on the several tens of femtosecond, resulting in a transient nonequilibrium occupation of phonons as well as a transient nonequilibrium carrier occupation near the Fermi level. The information about the incoming photon polarization is lost during this process. In the present work, the photoexcitation (pump) process is attributed to the DE because the high energy photoexcitation far above the band gap of CDW leads to the incoherent excitation. However, in the presence of a local, dynamic or hidden symmetry breaking, non-symmetric modes can be excited coherently even by the totally symmetric DE. In this case, the reflectivity change ΔR of the probe pulse can be divided into isotropic ΔR_{iso} and anisotropic ΔR_{aniso} components given by

$$\Delta R(\theta) \propto \Delta R_{iso} + \Delta R_{aniso} \cos (2\theta + \theta_0), \tag{3.36}$$

where θ is the polarization angle of the probe pulse and θ_0 is a phase offset, respectively. The polarization dependence of ΔR (i.e. ΔR_{aniso}) thus allows us to probe symmetry breaking in globally correlated electron systems.

The measurements were carried out using a two-color pump-probe setup, in which the coaxial configuration between the pump and probe beams allows to excite the precise position of the crystals even when using the vortex beam. This configuration is also advantageous for the polarization spectroscopy with regular reflection, in which the scattered pump was removed by the dichroic beam splitter. The light source used in this study was a mode-locked 100 fs Ti-sapphire laser oscillator operating at a repetition rate of about 76 MHz, which synchronously pumped an optical parametric oscillator (OPO). The probe pulse was extracted from the fundamental at $\sim$1.5 eV ($\sim$800 nm), and the pump was delivered by the OPO at $\sim$1.1 eV ($\sim$1100 nm). Both energies are much higher than the CDW gap energy in NbSe$_3$. We note that the probe energy that exceeds the pump energy allows to reduce the contributions from the higher excited states. Namely, this two-color combination enhances the optical response associated with the change of the conduction band-edge (CDW gap below T_c). Also note that transient signal was independent of the pump energy and polarization, which is responsible for the DE process mentioned above.

3.5.3 Pump-Probe Spectroscopy Using Polarization Vortex Pulses

Now we consider the pump-probe results for NbSe$_3$ samples (whisker and ring) using polarization vortex pulses with various azimuthal distributions [46]. In the experiment, the probe pulses are passed through a conversion system using an ASP, the details of which were described in Sect. 3.2.4. We note that all optical elements are achromatic in the near infrared region. The pump and probe beams were combined by a dichroic mirror and focused by using an achromatic objective lens. The focal spots of the two laser beams were adjusted to obtain a complete spatial overlap on the sample surface. Especially for the photoexcitation on the ring-shaped crystal, we carefully adjusted the vortex axis of the probe to place the center hole of the sample. The overlap between the two beams and the position on the sample surface were monitored using a CCD camera, and they were kept at a fixed position during the measurements. The samples were mounted on a cold finger of the helium-flow cryostat and all measurements are performed at ∼20 K.

Figure 3.16a–c shows transient ΔR measured by the probe pulse with various polarization distributions and angles. All data were plotted as a function of delay time between pump and probe pulses.

First, we focus on the result for the whisker sample using linearly polarized pulses (Fig. 3.16a). The polarization angle θ is defined from the b axis (chain axis) of the

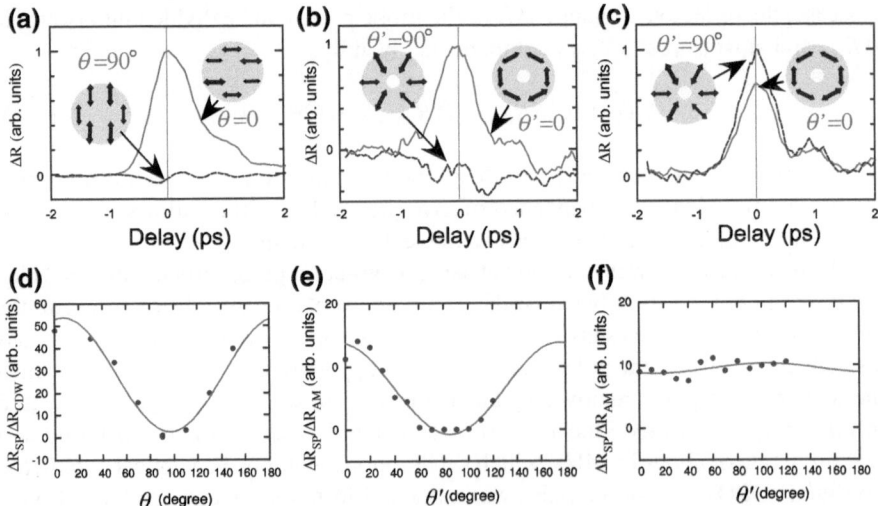

Fig. 3.16 Transient reflectivity changes ΔR for **a** a whisker crystal measured by the linear polarized pulse with $\theta = 0$ and 90°, **b** a (disk-like) ring crystal measured by the polarization vortex with $\theta' = 0$ and 90°, and **c** a whisker crystal measured by the polarization vortex with $\theta' = 0$ and 90°. Insets show the polarization distribution of the probe pulse. **d–f** show the corresponding polarization angle (θ or θ') dependence of the ΔR polarization in **a–c**, respectively, where the polarization is deduced from $\Delta R_{\mathrm{sp}}/\Delta R_{\mathrm{CDW}}$

whisker. As we mentioned in the previous subsection, the signal below T_c (below T_{c2}) is characterized by two features; exponential decay with a decay time of ~ 1 ps and a damped sinusoidal oscillation with a frequency of $\nu_{AM} = 1.12$ THz. We thus attribute the exponential part (ΔR_{sp}) to the carrier relaxation across the CDW gap and the oscillation part (ΔR_{CDW}) to the AM mode of the CDW. The exponential decay exhibits a significant polarization anisotropy, which can be associated with the 1D nature of the sample. The ΔR_{sp} was completely absent when the polarization is perpendicular to the b axis ($\theta = 90°$). In contrast, the oscillation component is nearly independent of the polarization θ, which is consistent with the fact that the AM is the fully symmetry A_1 mode. For the estimation of ΔR_{sp} and ΔR_{CDW}, we use the fitting function described by

$$\Delta R(t, \theta) = \Delta R_{sp}(\theta) \exp(-t/\tau_{sp}) + \Delta R_{CDW} \cos(2\pi \nu_{AM} t), \qquad (3.37)$$

where we neglect the damping of the AM that is much longer than the time scales observed here. The equation is compatible with (3.36), namely ΔR_{sp} is the anisotropic component $\Delta R_{aniso} \cos(2\theta)$ and ΔR_{CDW} corresponds to the isotropic component ΔR_{iso}.

Next, we consider the results for the ring sample using the polarization vortex pulses (Fig. 3.16b), where the θ' indicates the rotation angle of the local field polarization; $\theta' = 0°$ and $90°$ correspond to azimuthal and radial polarizations, respectively. By using such a topological polarization coordinate, the time evolution and its polarization dependence are related with those reported above in the whisker. Roughly speaking, Fig. 3.16b is in good agreement with Fig. 3.16a; the ΔR_{sp} component decreases with increasing θ' and vanishes at $\theta' = 90°$ while the oscillation component is constant with θ'. It is important to remember that the relative position between polarization vortex and center hole of the ring is critical for the ΔR polarization. We carefully adjusted the optical vortex onto the center hole of the ring.

In order to quantitatively compare the data of the ring and whisker, we plot the polarization of the ΔR_{sp} component as a function of θ' (θ) in Fig. 3.16d–f, where we estimate the degree of polarization from the ratio between ΔR_{sp} and ΔR_{CDW} in (3.37). Because of the fully symmetry of the AM oscillation, this method allows to cancel the remnant polarization of the experimental setup, thus being more accurate than the comparison of the absolute values of ΔR. The results in both data (Fig. 3.16d, e) clearly indicate the polarization of the ΔR_{sp} component whose magnitudes $\dfrac{I_{0°} - I_{90°}}{I_{0°} + I_{90°}}$ are near unity. Therefore we conclude that the ring sample exhibits a well-defined azimuthal symmetry responsible for the quasi 1D character. The polarization of the ring is also verified by the spatially-resolved ΔR detected with linearly polarized probe pulse (Fig. 3.17), where the magnitude of ΔR shows a bow-tie shaped distribution whose orientation is rotated by $90°$ between the two perpendicular probe polarizations. The results thus account for the azimuthal symmetry. However, by using the polarization vortices, we can analyze the global symmetry at once and quantitatively evaluate the degree of polarization. Moreover, our measurement also verifies that the CDWs in the ring maintain their 1D character, and

Fig. 3.17 Spatial maps of normalized ΔR obtained from the NbSe$_3$ (disk-like) ring sample. The geometry is consistent with Fig. 3.15d. The measurements were carried out with uniformly linear polarized probe pulse along **a** x and **b** y directions

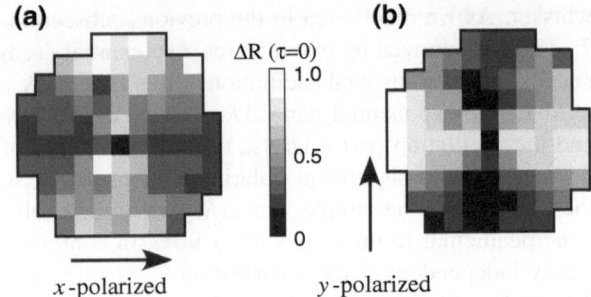

so that the technique using polarization vortex has an ability to evaluate the electron dynamics in closed-loop systems that plays an important role in the quantum correlation properties.

Finally, to enhance the accuracy of our measurement and analysis, we also show the results for the whisker using the polarization vortex pulses (Fig. 3.16c). In contrast to Fig. 3.16a, b, the ΔR in (c) exhibits no polarization. The constant SP polarization is also shown in Fig. 3.16f, where we can verify that the magnitudes of ΔR_{sp} are intermediate between the oscillatory magnitudes observed for the above two cases (Fig. 3.16d, e). This means that the polarization vortex with radial symmetry selects the local projection of the SP polarization along the azimuthally distributed θ'. Namely, the photoexcited 1D electrons act as a linear polarizer to evaluate the radial polarization distribution of the vortex pulses. The standard deviation of the magnitudes in Fig. 3.16f is 1.03, showing the high quality of the polarization vortex achieved in our time-resolved measurement.

3.6 Summary

This chapter has proposed a novel approach to ultrafast laser spectroscopy by means of topological light beams. Two kinds of spectroscopy have been presented: (1) nonlinear FWM spectroscopy using optical vortex (LG) pulses (Sect. 3.4), and (2) pump-probe reflection spectroscopy using polarization vortex pulses (Sect. 3.5). In short, the spatial characteristics of the topological light beams reveal unique properties of topologically-ordered electrons in each material system: excitons with OAM (Sect. 3.4) and closed-loop quasi-1D CDW (Sect. 3.5), each of which is summarized in detail below.

In Sect. 3.4, a coherent OAM transfer to the center of mass momentum of excitons in semiconductor GaN has been demonstrated using degenerate FWM process. The FWM signal excited with LG (optical vortex) pulses exhibits the phase singularities satisfying the conservation law of OAM. The results also demonstrated an arbitrary control of the topological charge in the FWM by changing the OAM of the excitation pulses, providing a way for controlling the optical OAM through carriers in semiconductor materials. Moreover, using an OAM-resolved detection system, we have realized to evaluate the dephasing dynamics of excitons topologically. We have observed the unfavorable $\ell = 0$ component associated with the imperfect phase conversion when one of the incident pulses includes $\ell = 0$ component. The effect was evidenced by the near perfect OAM conversion using the LG pulses with $\ell \neq 0$. In the latter case, the time-evolution of the OAM spectra exhibits a well-defined peak that remains almost unchanged during the dephasing time, clearly indicating that the center of mass momentum of excitons keep the OAM transferred by the LG pulses.

In Sect. 3.5, we technically realized an axisymmetrically-polarized vortex pulse excitation for a circulary closed quasi-1D CDW system in a two-color pump-probe setup. In a disk-like ring crystal, the transient reflectivity changes probed by the polarization vortex show a large anisotropy depending on the azimuthal polarization distributions, indicating that the quasi-1D electrons are polarized azimuthally and globally. These topologically-polarized electrons cannot be defined globally using the conventional photoexcitation (uniformly polarized optical pulses). Our demonstration provides a new spectroscopic technique for diagnosing the closed-loop coherence of carriers in solids, which should play an important role in the quantum correlation characteristics such as the Aharonov-Bohm effect.

Acknowledgments We are grateful to Prof. Yamane, Prof. Adachi, Prof. Oka, and Prof. Tanda (preparation of topological crystals in Sect. 3.5). We also wish to express our thanks to graduate and under-graduate students in our laboratory; Dr. Tokizane, Mr. Suzuki (Ph.D. student) and Mr. Shigematsu (PhD student).

References

1. L. Allen, M.W. Beijersbergen, R.J.C. Spreeuw, J.P. Woerdman, Orbital angular momentum of light and the transformation of Laguerre-Gaussian laser modes. Phys. Rev. A **45**, 8185–8189 (1992)
2. T. Ando, Y. Ohtake, N. Matsumoto, T. Inoue, N. Fukuchi, Mode purities of Laguerre-Gaussian beams generated via complex-amplitudemodulation using phase-only spatial light modulators. Opt. Lett. **34**, 34–36 (2009)
3. J.T. Barreiro, N.K. Langford, N.A. Peters, P.G. Kwiat, Generation of hyperentangled photon pairs. Phys. Rev. Lett. **95**, 260501 (2005)
4. M.W. Beijersbergen, L. Allen, H.E.L.O. van der Veen, J.P. Woerdman, Astigmatic laser mode converters and transfer of orbital angular momentum. Opt. Commun. **96**, 123–132 (1993)
5. M.W. Beijersbergen, R.P.C. Coerwinkel, M. Kristensen, J.P. Woerdman, Helical-wavefront laser beams produced with a spiral phaseplate. Opt. Commun. **112**, 321–327 (1994)
6. M.V. Berry, M.R. Dennis, Jr. R L. Lee, Polarization singularities in the clear sky. New J. Phys. **6**, 162 (2004)
7. Z. Bouchal, R. Celechovsky Mixed vortex states of light as information carriers. New J. Phys. **6**, 131 (2004)
8. N. Bozinovic, Y. Yue, Y. Ren, M. Tur, P. Kristensen, H. Huang, A.E. Willner, S. Ramachandran, Terabit-scale orbital angular momentum mode division multiplexing in fibers. Science **340**, 1545–1548 (2013)
9. M. Brambilla, F. Battipede, L.A. Lugiato, V. Penna, F. Prati, C. Tamm, C.O. Weiss, Transverse laser patterns. i. phase singularity crystals. Phys. Rev. A **43**, 5090–5113 (1991)

10. S. Bretschneider, C. Eggeling, S.W. Hell, Breaking the diffraction barrier in fluorescence microscopy by optical shelving. Phys. Rev. Lett. **98**, 218103 (2007)
11. A.V. Carpentier, H. Michinel, J.R. Salgueiro, D. Olivieri, Making optical vortices with computer-generated holograms. Am J Phys. **76**, 916–921 (2008)
12. I. Carusotto, C. Ciuti, Quantum fluids of light. Rev. Mod. Phys. **85**, 299–366 (2013)
13. R. Celechovsky, Z. Bouchal, Optical implementation of the vortex information channel. New J. Phys. **9**, 328 (2007)
14. M.A. Cibula, D.H. McIntyre, General algorithm to optimize the diffraction efficiency of a phase-type spatial light modulator. Opt. Lett. **38**, 2767–2769 (2013)
15. J. Demsar, K. Biljakovic, D. Mihailovic, Single particle and collective excitations in the one-dimensional charge density wave solid k0.3moo3 probed in real time by femtosecond spectroscopy. Phys. Rev. Lett. **83**, 800–803 (1999)
16. V. Denisenko, V. Shvedov, A.S. Desyatnikov, D.N. Neshev, W. Krolikowski, A. Volyar, M. Soskin, Y.S. Kivshar, Determination of topological charges of polychromatic optical vortices. Opt. Express **17**, 23374–23379 (2009)
17. S. Franke-Arnold, L. Allen, M. Padgett, Advances in optical angular momentum. Laser Photonics Rev. **2**, 299–313 (2008)
18. G. Gibson, J. Courtial, M.J. Padgett, M. Vasnetsov, V. Pas'ko, S.M. Barnett, S. Franke-Arnold, Free-space information transfer using light beams carrying orbital angular momentum. Opt. Express **12**, 5448–5456 (2004)
19. J. Hamazaki, R. Morita, K. Chujo, Y. Kobayashi, S. Tanda, T. Omatsu, Optical-vortex laser ablation. Opt. Express **18**, 2144–2151 (2010)
20. M. Harris, C.A. Hill, P.R. Tapster, J.M. Vaughan, Laser modes with helical wave fronts. Phys. Rev. A **49**, 3119–3122 (1994)
21. N.R. Heckenberg, R. McDuff, C.P. Smith, A.G. White, Generation of optical-phase singularities by computer-generated holograms. Opt. Lett. **17**, 221–223 (1992)
22. S.W. Hell, Toward fluorescence nanoscopy. Nat. Biotechnol. **21**, 1347–1355 (2003)
23. G. Indebetouw, Optical vortices and their propagation. J. Mod. Opt. **40**, 73–87 (1993)
24. J. Jimenez, Y. Noblet, P.V. Paulau, D. Gomila, T. Ackemann, Observation of laser vortex solitons in a self-focusing semiconductor laser. J. Opt. **15**, 044011 (2013)
25. Y. Keisaku, Y. Zhili, T. Yasunori, M. Ryuji, Frequency-resolved measurement of the orbital angular momentum spectrum of femtosecond ultra-broadband optical-vortex pulses based on field reconstruction. New J. Phys. **16**, 053020 (2014)
26. J.W. Kim, J.I. Mackenzie, J.R. Hayes, W.A. Clarkson, High power er:yag laser with radially-polarized Laguerre-Gaussian (lg01) mode output. Opt. Express **19**, 14526–14531 (2011)
27. M. Koyama, T. Hirose, M. Okida, K. Miyamoto, T. Omatsu, Nanosecond vortex laser pulses with millijoule pulse energies from a yb-doped double-clad fiber power amplifier. Opt. Express **19**, 14420–14425 (2011)
28. S. Kyohhei, T. Yasunori, Y. Keisaku, M. Ryuji, Orbital angular momentum spectral dynamics of gan excitons excited by optical vortices. Jpn. J. Appl. Phys. **52**, 08JL08 (2013)
29. J. Leach, M.R. Dennis, J. Courtial, M.J. Padgett, Vortex knots in light. New J. Phys. **7**, 55 (2005)
30. L.A. Lugiato, F. Prati, L.M. Narducci, G.L. Oppo, Spontaneous breaking of the cylindrical symmetry in lasers. Opt. Commun. **69**, 387–392 (1989)
31. A. Mair, A. Vaziri, G. Weihs, A. Zeilinger, Entanglement of the orbital angular momentum states of photons. Nature **412**, 313–316 (2001)
32. T. Matsuura, K. Inagaki, S. Tanda, Evidence of circulating charge density wave current: shapiro interference in nbse3 topological crystals. Phys. Rev. B **79**, 014304 (2009)
33. P. Monceau, Electronic crystals: an experimental overview. Adv. Phys. **61**, 325–581 (2012)
34. Y. Mushiake, K. Matsumura, N. Nakajima, Generation of radially polarized optical beam mode by laser oscillation. Proc. IEEE. **60**, 1107–1109 (1972)
35. J.F. Nye, M.V. Berry, Dislocations in wave trains. Proc. R. Soc. Lond. A: Math. Phys. Sci. **336**, 165–190 (1974)

36. M.J. Padgett, L. Allen, The poynting vector in Laguerre-Gaussian laser modes. Opt. Commun. **121**, 36–40 (1995)
37. D. Pohl, Operation of a ruby laser in the purely transverse electric mode TE 01. Appl. Phys. Lett. **20**, 266–267 (1972)
38. D. Rozas, Z.S. Sacks, G.A. Swartzlander, Experimental observation of fluidlike motion of optical vortices. Phys. Rev. Lett. **79**, 3399–3402 (1997)
39. J. Sato, M. Endo, S. Yamaguchi, K. Nanri, T. Fujioka, Simple annular-beam generator with a laser-diode-pumped axially off-set power build-up cavity. Opt. Commun. **277**, 342–348 (2007)
40. J. Shah, *Ultrafast Spectroscopy of Semiconductors and Semiconductor Nanostructures*, vol. 115. (Springer, Cambridge 1999)
41. K. Shimatake, Y. Toda, S. Tanda, Selective optical probing of the charge-density-wave phases in nbse3. Phys. Rev. B **75**, 115120 (2007)
42. M. Stalder, M. Schadt, Linearly polarized light with axial symmetry generated by liquid-crystal polarization converters. Opt. Lett. **21**, 1948–1950 (1996)
43. S. Tanda, T. Tsuneta, Y. Okajima, K. Inagaki, K. Yamaya, N. Hatakenaka, Crystal topology: a möbius strip of single crystals. Nature **417**, 397–398 (2002)
44. Y. Toda, K. Shigematsu, K. Yamane, R. Morita, Efficient Laguerre-Gaussian mode conversion for orbital angular momentum resolved spectroscopy. Opt. Commun. **308**, 147–151 (2013)
45. Y. Tokizane, K. Oka, R. Morita, Supercontinuum optical vortex pulse generation without spatial ortopological-charge dispersion. Opt. Express **17**, 14517–14525 (2009)
46. Y. Tokizane, K. Shimatake, Y. Toda, K. Oka, M. Tsubota, S. Tanda, R. Morita, Global evaluation of closed-loop electron dynamics in quasi-one-dimensional conductors using polarization vortices. Opt. Express **17**, 24198–24207 (2009)
47. L. Torner, J.P. Torres, S. Carrasco, Digital spiral imaging. Opt. Express **13**, 873–881 (2005)
48. M. Tsubota, K. Inagaki, T. Matsuura, S. Tanda, Aharonov-bohm effect in charge-density wave loops with inherent temporal current switching. Europhys. Lett. **97**, 57011 (2012)
49. R.K. Tyson, M. Scipioni, J. Viegas, Generation of an optical vortex with a segmented deformable mirror. Appl. Opt. **47**, 6300–6306 (2008)
50. Y. Ueno, Y. Toda, S. Adachi, R. Morita, T. Tawara, Coherent transfer of orbital angular momentum to excitons by opticalfour-wave mixing. Opt. Express **17**, 20567–20574 (2009)
51. J. Wang, J.-Y. Yang, I.M. Fazal, N. Ahmed, Y. Yan, H. Huang, Y. Ren, Y. Yue, S. Dolinar, M. Tur, A.E. Willner, Terabit free-space data transmission employing orbital angular momentum multiplexing. Nat. Photonics **6**, 488–496 (2012)
52. K. Yamane, Y. Toda, R. Morita, Ultrashort optical-vortex pulse generation in few-cycle regime. Opt. Express **20**, 18986–18993 (2012)
53. Y. Yoshikawa, H. Sasada, Versatile generation of optical vortices based on paraxial mode expansion. J. Opt. Soc. Am. A **19**, 2127–2133 (2002)
54. K. Youngworth, T. Brown, Focusing of high numerical aperture cylindrical-vector beams. Opt. Express **7**, 77–87 (2000)
55. Q. Zhan, Cylindrical vector beams: from mathematical concepts to applications. Adv. Opt. Photon. **1**, 1–57 (2009)

Chapter 4
Localized Modes in Nonlinear Discrete Systems

Kazuyuki Yoshimura, Yusuke Doi and Masayuki Kimura

Abstract This chapter reviews spatially localized modes emerging in nonlinear discrete dynamical systems, which are called intrinsic localized modes. After the notion of intrinsic localized mode is introduced by using a simple mathematical model, the essential mechanism for the existence of intrinsic localized modes and their basic properties are described in both cases of standing wave and traveling wave modes. The intrinsic localized modes in real systems have also been studied so far. We review some numerical simulation studies and experimental observations of the intrinsic localized modes in real systems.

4.1 Introduction

A variety of spatially extended systems in nature have both spatially-discrete structures and nonlinearity. Such systems are as diverse as solid crystals, polymer molecules, mechanical or electrical devices, and photonic crystals. Those systems can be mathematically modeled by nonlinear space-discrete dynamical systems. The ground-breaking work by Takeno et al.[1, 2] has theoretically revealed that spatially localized excitations are ubiquitously possible in dynamical systems with discreteness and nonlinearity. The localized modes are called *intrinsic localized modes* (ILMs) or *discrete breathers*. A common expectation for a spatially extended system may be that an initially localized excitation energy will spread over the system in

K. Yoshimura (✉)
NTT Communication Science Laboratories, NTT Corporation, 2-4, Hikaridai, Seika-cho, Sorakugun, Kyoto 619-0237, Japan
e-mail: yoshimura.kazuyuki@lab.ntt.co.jp

Y. Doi
Department of Adaptive Machine Systems, Graduate School of Engineering, Osaka University, 2-1, Yamadaoka, Suita, Osaka 565-0871, Japan
e-mail: doi@ams.eng.osaka-u.ac.jp

M. Kimura
Department of Electric Engineering, Kyoto University, A1-418, Katsura, Nishikyo-ku, Kyoto 615-8510, Japan
e-mail: kimura.masayuki.8c@kyoto-u.ac.jp

© Springer International Publishing Switzerland 2015
M. Ohtsu and T. Yatsui (eds.), *Progress in Nanophotonics 3*,
Nano-Optics and Nanophotonics, DOI 10.1007/978-3-319-11602-0_4

the course of time. Contrary to the expectation, the existence of ILM implies that it prevents such delocalization of the energy, retaining a localized excitation structure. Only the properties of a dynamical system necessary for the existence of ILM are discreteness and nonlinearity, which are possessed by many systems. This fact ensures the universality of ILM in model and real systems. Indeed, there are a large number of theoretical and experimental evidence for the existence of ILM in various systems: for instance, the ILMs have been experimentally observed in Josephson junction arrays [3, 4], optical waveguide arrays [5], and micromechanical systems [6, 7] (see also [8–11] and references therein). The concept of ILM has attracted great interest because of its universality.

It is expected that the concept of ILM is useful to understand essential mechanisms of energy localization, energy transport, and energy exchange processes and of related phenomena in various systems in nature, although unfortunately the role of ILM in nature has not yet been clarified. The concept of ILM is expected to be useful also in the field of nanophotonics. A recent molecular dynamics study has shown that a local structural transition in a carbon nanotube, i.e., only a few atoms change their configurations, can be induced by energy localization due to ILM [12]. A similar local structural transition is known in a photoinduced phase transition material, which is used for optical memory media and records digital information by using this transition. The mechanism of this transition has not yet been clarified. The concept of ILM could explain the mechanism. Another example is the *dressed photon phonon* (DPP). The dressed photon (DP) is a quasi-particle composed of photons and an electron-hole pair [13]. Recent experiments have shown that there are some phenomena which cannot be explained by the DP concept, and they indicated necessity of considering a quasi-particle composed of DP and phonons. This quasi-particle is the DPP. Theoretical study by Ohtsu has shown that a DP and a localized phonon mode can couple with each other to form the DPP quasi-particle [13]. In their study, a linear lattice with impurities was used and the localized phonon is just due to the impurities. If the lattice is a nonlinear one, there is a localized phonon mode, i.e., ILM, and it could couple with a DP to form a *nonlinear* DPP. It may be worth exploring the possibility of nonlinear DPP. As the above examples suggest, there may be close connections between the ILM concept and problems in nanophotonics.

This review article aims at providing an introduction of the ILM to expedite its applications in nanophotonics. Fundamental notions of the ILM and recent progress in the field are described. This article is organized as follows. In Sect. 4.2, we introduce a mathematical model for studying ILM, i.e., a lattice model, and explain what the ILM is. The essential mechanism for the existence of ILM is also explained. There are two types of ILMs, which are of standing wave type and traveling wave type. We focus on the standing wave ILM in Sects. 4.3 and 4.4 while on the traveling wave one in Sect. 4.5. In Sect. 4.3, we describe some fundamental properties of the standing wave ILM based on a simple lattice model. In addition, some known mathematically rigorous results are mentioned. In Sect. 4.4, we introduce several lattice models and show that different types of standing wave ILMs are possible in them. In Sect. 4.5, we describe fundamental properties of the traveling wave ILMs, including their collision dynamics. The rest sections deal with ILMs in real systems. We present numerical simulation and experimental results on the ILM dynamics in real systems in Sects. 4.6 and 4.7, respectively.

4.2 What Is the ILM?

4.2.1 Lattice Model

Lattice is a simple mathematical model, which is used for studying fundamental dynamical properties of real systems having spatially discrete structures. The lattice model describes the dynamics of many interacting particles. For simplicity, in this review article, we mainly address one-dimensional lattices which are described by the Hamiltonian

$$H = \sum_n \frac{1}{2m_n} p_n^2 + \sum_n \left[U(q_n) + V(q_{n+1} - q_n) \right], \tag{4.1}$$

where $q_n \in \mathbb{R}$, $p_n \in \mathbb{R}$, and m_n represent the position, momentum, and mass of nth particle, respectively. Hamiltonian (4.1) describes a one-dimensional chain of particles such that each particle interacts with its nearest neighbours via springs and oscillates in an on-site potential. Figure 4.1 illustrates this one-dimensional chain of particles. The functions $U(X) \colon \mathbb{R} \to \mathbb{R}$ and $V(X) \colon \mathbb{R} \to \mathbb{R}$ represent the on-site and interaction potential functions, respectively. We assume they are smooth functions.

Hamilton's equations of motion are given by

$$\dot{q}_n = \frac{\partial H}{\partial p_n}, \quad \dot{p}_n = -\frac{\partial H}{\partial q_n}. \tag{4.2}$$

For lattice Hamiltonian (4.1), these equations can be written in the form

$$m_n \ddot{q}_n = -U'(q_n) + V'(q_{n+1} - q_n) - V'(q_n - q_{n-1}). \tag{4.3}$$

We did not specify the boundary conditions in (4.3). Boundary conditions often used in literature are fixed-end conditions and periodic conditions. The fixed-end conditions are given by $q_0(t) = q_{N+1}(t) = 0$ while the periodic ones are given

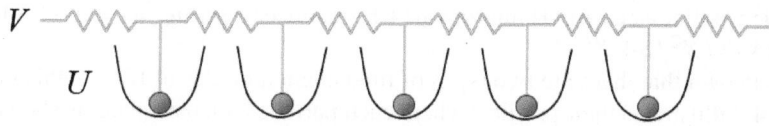

Fig. 4.1 Illustration of one-dimensional lattice

by $q_0(t) = q_N(t)$ and $q_{N+1}(t) = q_1(t)$, provided that $n = 1, 2, \ldots, N$ in (4.3). In addition, infinite-size lattices are often used, which have no boundary condition. It depends on the problem which type of boundary conditions is employed.

There are various choices of the potential functions U and V and the set $\{m_n\}$ of masses. As for the masses, the simplest case is that all the particles have identical masses, i.e., $m_n = m$ for all n, where m is a positive constant. Hereafter, we assume this case for the sake of simplicity. A model of identical masses is sometimes called a *monoatomic lattice*. Various types of functions are possible for the potentials U and V, and some lattices are particularly named. We present a few examples below.

Example 1. Consider the case of harmonic potentials, i.e., $U(X) = \mu X^2$ and $V(X) = \kappa X^2$, where $\mu \geq 0$ and $\kappa \geq 0$ are constants. This model is called a harmonic lattice or a linear lattice.

Example 2. Consider the case that $U(X) = 0$ and $V(X) = \kappa_2 X^2 + \kappa_3 X^3$ or $\kappa_2 X^2 + \kappa_4 X^4$, where κ_j are real constants. These models are called Fermi-Pasta-Ulam (FPU) lattices as they were used in the seminal paper by these authors [14], which reported the first numerical experiment results on the ergodic problem. More precisely, the cubic potential model is called the FPU-α lattice, and the quartic potential one the FPU-β lattice. In some literature, lattices with more general anharmonic functions V are also called by the same name.

Example 3. Consider the case that $U(X)$ is anharmonic and $V(X) = \kappa_2 X^2$. This model is called a nonlinear Klein–Gordon (NKG) lattice.

We mainly deal with the case of Hamiltonian lattices in the following sections, since mathematical properties of ILM in Hamiltonian lattices are of the most fundamental. Non-Hamiltonian lattice models can also be considered, which have dissipation and external forces. It is relatively easy to include effects of dissipation and forcing.

4.2.2 Fundamental Modes of ILM

Let us consider a FPU lattice which has an interaction potential function consisting of quadratic and quartic terms, i.e., the FPU-β lattice, as a simple example of nonlinear lattice. The Hamiltonian is given by

$$H = \sum_n \frac{1}{2} p_n^2 + \sum_n \left[\frac{1}{2}(q_{n+1} - q_n)^2 + \frac{1}{4}(q_{n+1} - q_n)^4 \right], \qquad (4.4)$$

which is a particular case of Hamiltonian (4.1) obtained by setting $m_n = 1$, $U(X) = 0$, and $V(X) = X^2/2 + X^4/4$.

It is known that there are two types of fundamental modes of ILM in this model. Figure 4.2 illustrates their profiles, where each particle's displacement is shown in a vertical direction and arrows represent velocity vectors of the particles. Two neighbouring particles periodically oscillate anti-phase in both modes, as shown by the

(a) **(b)**

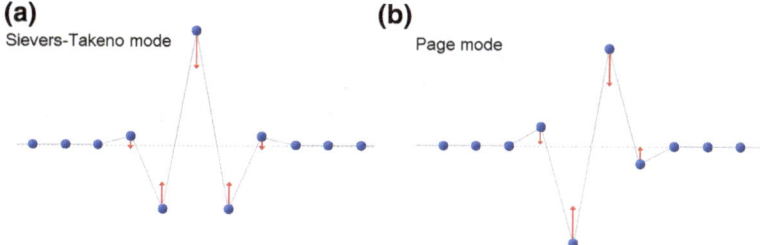

Sievers-Takeno mode   Page mode

Fig. 4.2 Profile of ILM. **a** Sievers–Takeno mode and **b** Page mode

arrows. These two modes have different spatial symmetries in their profiles. One mode has its center of profile at a site while the other mode at the midpoint between two sites, i.e., site-centered and bond-centered modes. The former is called Sievers–Takeno (ST) mode or odd mode, and the latter is called Page (P) mode or even mode. Sievers and Takeno showed that the ST mode is possible to exist in the FPU lattice (4.4) in [1, 2]. Two years later, Page found the P mode [15].

The amplitude of each particle rapidly decreases as the distance from the mode center increases in both of the ST and P modes. In this sense, these two modes are spatially localized. Analytical calculations by Sievers and Takeno and by Page showed that the normalized spatial profiles of ST and P modes are approximately given by $(\dots, 0, -1/2, 1, -1/2, 0, \dots)$ and $(\dots, 0, 1, -1, 0, \dots)$ in the limit of strong localization, respectively. As we will discuss in Sect. 4.3, an ILM comes to localize more strongly as its frequency increases. Therefore, the strong localization limit corresponds to the limit of large frequency.

In mathematical terms, ILMs are time-periodic and spatially localized solutions of the equations of motion. An ILM is a solution satisfying some sort of localization conditions and the periodicity conditions

$$q_n(t + T) = q_n(t), \quad p_n(t + T) = p_n(t), \tag{4.5}$$

where T is the period of ILM. The ILM frequency is defined by $\Omega = 2\pi/T$. A commonly-accepted definition of localization conditions has not been given definitively. There is some arbitrariness in choosing the conditions, depending on the problem. For example, in the case of infinite-size lattices with U and V even functions, a set of natural conditions may be

$$\max_{0 \le t \le T} |q_n(t)| \to 0, \quad n \to \pm\infty. \tag{4.6}$$

In the case of finite-size lattices, a natural definition is not apparent. It is necessary to introduce an appropriate definition suitable for the problem. Because of the above mentioned arbitrariness, the term "localization" should be understood in a *physical* sense, instead of a mathematical sense.

The ST and P mode solutions were originally found by approximate analytical calculations and then numerically confirmed in several works (e.g., [16]). In addition, some mathematical proofs have already been given for the existence of exact periodic solutions corresponding to these modes. The issue of mathematically rigorous results will be addressed in Sect. 4.3.4.

4.2.3 Numerical Solutions of ILM

Let us consider a finite-size FPU lattice (4.4), for which $n = 1, 2, \ldots, N$ and the fixed-end boundary conditions $q_0(t) = q_{N+1}(t) = 0$ are assumed. An ILM solution corresponds to a closed orbit in phase space $\mathbb{R}^{2N}$ spanned by the coordinates $\{q_n, p_n\}$. It is known that ILM solutions for different energy values form a one-parameter family of closed orbits parametrized by energy, which can be also parametrized by period, frequency or amplitude of the ILMs instead of energy. Given a period T, it is possible to compute a closed orbit corresponding to the ILM solution with period T by using a Newton method. This numerical method is described in Appendix.

Figure 4.3 shows examples of ILM solutions obtained numerically. The particle displacements q_n are shown for the initial time when velocities of all the particles

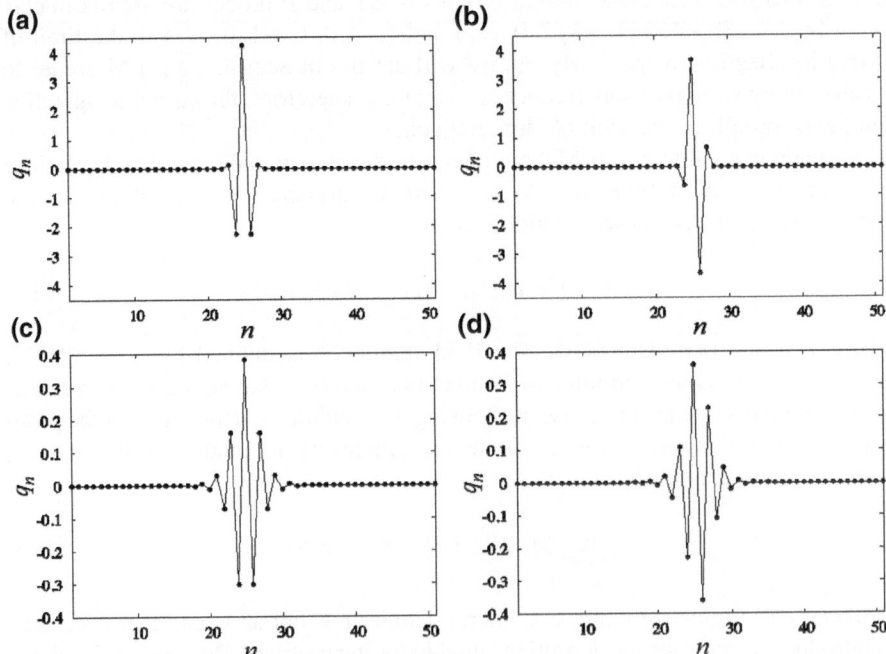

Fig. 4.3 Numerical solution of ILM in FPU lattice (4.4). The lattice size is $N = 51$, and fixed-end boundary conditions are used. Displacement $q_n(0)$ is plotted against site number n for **a** ST mode with $\Omega = 10$, **b** P mode with $\Omega = 10$, **c** ST mode with $\Omega = 2.2$, and **d** P mode with $\Omega = 2.2$

vanish. ST and P mode solutions are shown for two different frequencies $\Omega = 2.2$ and 10. The larger frequency corresponds to a larger energy case. Mode profiles close to those in Fig. 4.2 are confirmed for large Ω. In contrast, for small Ω, the solutions are only weakly localized. This shows that ILM comes to localize more strongly as its frequency increases. It is generally observed that spatial decrease of particles' displacements is exponential, provided that the lattice has a harmonic term in its interaction potential. It is known that super exponential decrease is observed when the interaction potential has no harmonic term [17].

4.2.4 Necessity of Nonlinearity for ILM

We emphasize that the localized modes, ILMs, are peculiar to nonlinear lattices and they are not expected in linear lattices, provided that the lattices are monoatomic. To show this, we compare behaviors of a nonlinear lattice and a linear one numerically. We use the FPU lattice (4.4) as an example of nonlinear lattice. For comparison, we use a linear lattice described by the Hamiltonian

$$H_L = \sum_n \frac{1}{2} p_n^2 + \sum_n \frac{1}{2}(q_{n+1} - q_n)^2, \tag{4.7}$$

which is obtained by setting $m_n = 1$, $U(X) = 0$, and $V(X) = X^2/2$ in (4.1). We assume $n = 1, 2, \ldots, N$, and the fixed-end boundary conditions $q_0(t) = q_{N+1}(t) = 0$ are used in (4.4) and (4.7).

Let us consider a locally excited initial condition such that $q_{(N+1)/2}(0) = 1$, $q_n(0) = 0$ for $n \neq (N+1)/2$, and $p_n(0) = 0$. We numerically integrated the equations of motion of the two lattices with this initial condition. To show the numerical results, we define the energy e_i of ith site by the sum of the kinetic energy of the ith particle and the half of the interaction potential energies shared with its two nearest neighbors. The site energy e_i, $i = 1, 2, \ldots, N$ is given by

$$e_n = \frac{1}{2} p_n^2 + \frac{1}{2} \left[V(q_{n+1} - q_n) + V(q_n - q_{n-1}) \right], \tag{4.8}$$

where V is the interaction potential of the FPU or linear lattice.

Figure 4.4a, b shows time evolutions of site energy e_n for the FPU and linear lattices, respectively, where e_n is shown with different colors as a function of site number n and time t. Figure 4.4a clearly shows that a stable localized mode appears after an initial radiation of small ripples, and it remains for a long time in the FPU lattice. Note that this long-lasting localized mode emerges from an initial condition which differs from an exact ILM solution. This fact suggests that a localized excitation emerges robustly for a wide class of initial conditions. In contrast, an initially localized structure becomes delocalized and eventually only a spatially extended

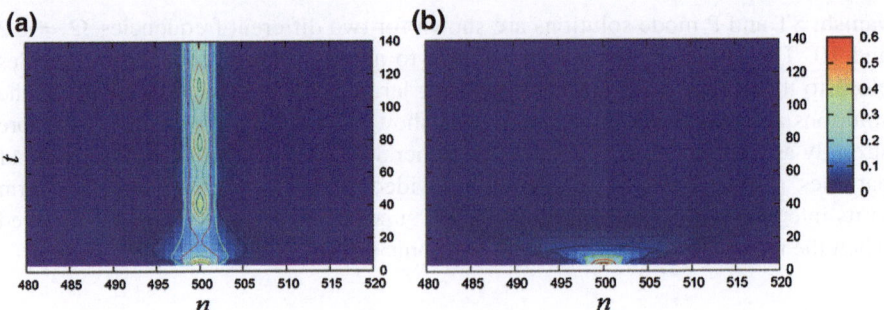

Fig. 4.4 Time evolution of site energy in **a** FPU lattice (4.4) and **b** linear lattice. The lattice size is $N = 999$

state emerges in the linear lattice. The above comparison clearly demonstrates that localized modes are peculiar to nonlinear lattices.

4.2.5 Mechanism for Existence of ILMs

ILMs are exact periodic solutions of lattices, which can exist due to discreteness and nonlinearity of the lattices. The periodicity of an ILM implies that its localized structure is retained without any decay during the time evolution. We use the FPU lattice (4.4) as an example of nonlinear lattice and intuitively elucidate the essential mechanism that makes it possible to retain the localized structure, focusing on roles of discreteness and nonlinearity.

First, we explain a role of discreteness, based on a difference between a spatially discrete medium, i.e., a lattice, and a spatially continuous medium. We do not consider any nonlinearity here since we focus on the role of discreteness only. Let us consider a one-dimensional continuous medium described by the partial differential equation

$$\frac{\partial^2 u}{\partial t^2} = c^2 \frac{\partial^2 u}{\partial x^2},$$

(4.9)

where $c > 0$ is a constant. If we substitute a plane wave solution $u(t, x) = \exp[i(kx - \omega t)]$ into (4.9), then we obtain the dispersion relation

$$\omega^2 = c^2 k^2.$$

(4.10)

This relation indicates that for any value of frequency $\omega \geq 0$ there exists a plane wave solution of (4.9) with the ω and a wavenumber k determined by (4.10).

Let us move to a linear lattice defined by Hamiltonian (4.7). The equations of motion are given by

$$\ddot{q}_n = q_{n+1} - 2q_n + q_{n-1}.$$

(4.11)

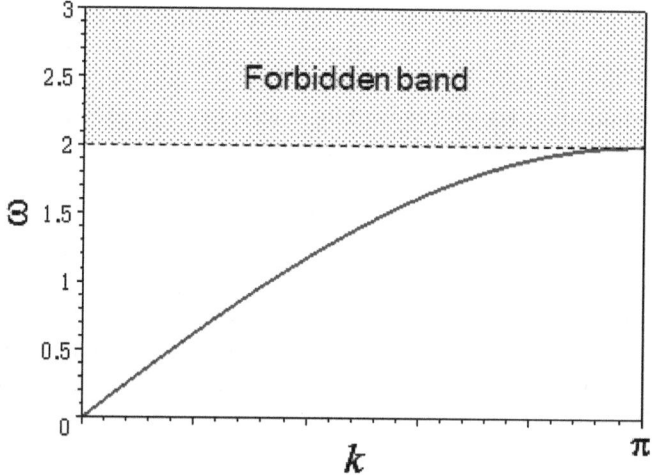

Fig. 4.5 Dispersion relation of lattice (4.7)

Substitution of $q_n(t) = \exp[i(kn - \omega t)]$ into (4.11) yields the dispersion relation

$$\omega^2 = 4\sin^2(k/2). \tag{4.12}$$

The range of k is limited to $-\pi \le k \le \pi$ because of spatial discreteness of the lattice. This dispersion relation is shown in Fig. 4.5. An important feature of (4.12) is that there exists the maximum value $\omega_{max} = 2$ in frequency, which is qualitatively different from (4.10). This implies that a plane wave solution cannot exist in the frequency range $\omega > 2$, which is called a *forbidden band*. In this connection, the range $0 \le \omega \le \omega_{max}$ is called a *phonon band*. We emphasize that the existence of a forbidden band is a peculiarity due to spatial discreteness of the lattice.

To see a consequence of this peculiarity, let us consider a semi-infinite linear lattice with $n = 0, 1, 2, \ldots$, which is described by (4.11) and imposed with the time-dependent boundary condition $q_0(t) = A \exp[i\omega t]$, where A is a constant. We explicitly construct a periodic solution of this system, assuming the form $q_n(t) = u_n \exp[i\omega t]$, where u_n is a constant. Substituting this form into (4.11), we have the difference equation

$$u_{n+1} - \left(2 - \omega^2\right) u_n + u_{n-1} = 0. \tag{4.13}$$

This equation can be solved by assuming the ansatz $u_n = Ar^n$. This ansatz yields

$$r^2 - \left(2 - \omega^2\right) r + 1 = 0. \tag{4.14}$$

Solutions of this quadratic equation are qualitatively different between the two cases (i) $0 \leq \omega \leq \omega_{max}$ and (ii) $\omega_{max} < \omega$, where $\omega_{max} = 2$. In case (i), we have solutions of the form $r = \exp[\pm i\theta]$, $\theta \in \mathbb{R}$. These solutions lead to the spatially-extended plane wave solutions $q_n(t) = A \exp[i(\pm \theta n + \omega t)]$ of (4.11). On the other hand, in case (ii), solutions of (4.14) are obtained as

$$r = 1 - \frac{\omega}{2} \left[\omega \pm \sqrt{\omega^2 - 4} \right]. \tag{4.15}$$

The solution of negative sign satisfies $-1 < r < 0$, and only this solution is physically relevant since u_n diverges as $n \to \infty$ for the other one. The corresponding solution $q_n(t) = A r^n \exp[i\omega t]$ of (4.11) exhibits anti-phase oscillation between two neighbouring particles, and it is localized in the sense that its amplitude decreases exponentially with respect to n. This solution exists for any ω in the forbidden band.

Second, we explain a role of nonlinearity. Consider an infinite FPU lattice defined by Hamiltonian (4.4). Assume that an exact ILM state is realized at the initial, which is illustrated in Fig. 4.6. Particles of core sites with $|n| \leq n_0$ oscillate periodically with frequency Ω. The role of nonlinearity of the FPU potential is to induce a value of Ω in the forbidden band. Note that $\Omega > \omega_{max}$ is impossible without nonlinearity. Let us consider two lattice segments outside the core part, i.e., $|n| \geq n_0$, where the harmonic term in potential may be dominant because of small amplitudes of particles. These segments are substantially driven by the boundary sites of $n = \pm n_0$ with $\Omega > \omega_{max}$. The previous argument ensures that periodic oscillations with exponentially decaying amplitude with $n \to \pm\infty$ realize and continue there. This is the reason why the initial localized state is retained. In contrast, the initial localized state cannot be retained but it collapses if the potential is harmonic, because $\Omega \leq \omega_{max}$ and then the exponentially decaying profiles cannot last in the side segments $|n| \geq n_0$.

To summarize, the spatial discreteness causes the forbidden band and the nonlinearity makes it possible to achieve a frequency in the forbidden band. A combined effect of these two factors gives the essential mechanism which enables ILMs to keep their localization profiles in the time evolution.

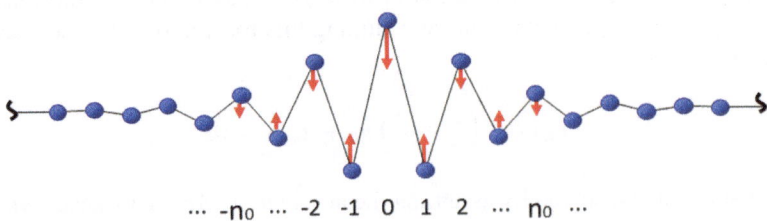

Fig. 4.6 Illustration of an initial state given by ILM

4.2.6 Fundamental Modes in Other Lattices

The mechanism described in Sect. 4.2.5 does not require any special mathematical property of a lattice such as integrability. Therefore, it applies to a variety of nonlinear lattices other than the FPU lattice (4.4). This fact supports universality of ILM. In this subsection, we mention about some examples of ILMs in other lattices.

Let us consider the NKG lattice defined by the Hamiltonian

$$H = \sum_n \frac{1}{2}p_n^2 + \sum_n \left[U(q_n) + \frac{\kappa}{2}(q_{n+1} - q_n)^2 \right],$$ (4.16)

where $U(X) = X^2 + X^3 + X^4/4$ and $\kappa = 0.1$. This example is borrowed from [11]. The dispersion relation is

$$\omega^2 = 2 + 4\kappa \sin^2(k/2),$$ (4.17)

shown in Fig. 4.7, and it has both minimum and maximum frequencies. This model has two forbidden bands $0 \leq \omega < \omega_{min}$ and $\omega_{max} < \omega$. ILMs can have their frequencies in either of these forbidden bands, based on the mechanism in Sect. 4.2.5. Figure 4.8 shows numerical ILM solutions for two different frequencies located in the upper and lower forbidden bands. These numerical solutions were computed by using a method in Appendix. For each frequency, we show two ILMs of site-centered and bond-centered profiles. The high-frequency ILMs ($\Omega = 1.7$) exhibit motions with adjacent particles moving anti-phase, which are indicated by a staggered displacement pattern in Fig. 4.8a, b. In contrast, non-staggered patterns in Fig. 4.8c,

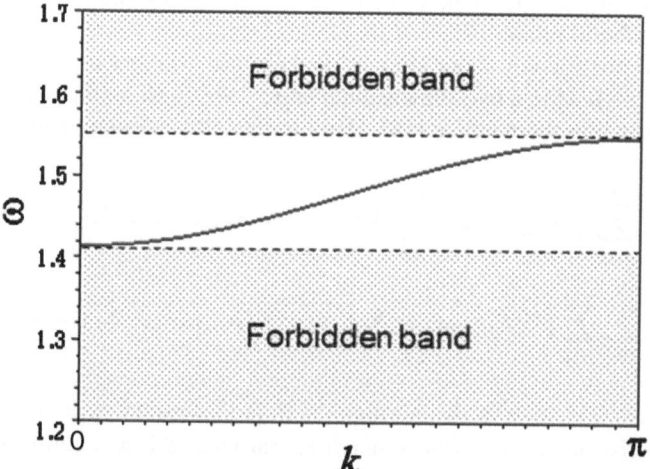

Fig. 4.7 Dispersion relation of NKG lattice (4.16)

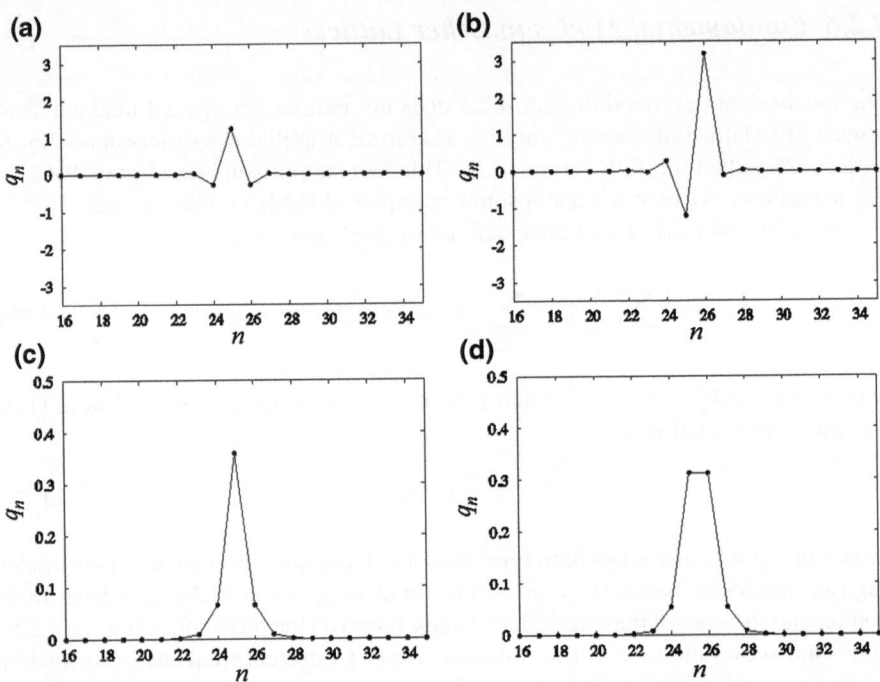

Fig. 4.8 Numerical solution of ILM in NKG lattice (4.16). The lattice size is $N = 51$, and fixed-end boundary conditions are used. Displacement $q_n(0)$ is plotted against site number n for **a** site-centered mode with $\Omega = 1.7$, **b** bond-centered mode with $\Omega = 1.7$, **c** site-centered mode with $\Omega = 0.8$, and **d** bond-centered mode with $\Omega = 0.8$

d indicate that the low-frequency ILMs ($\Omega = 0.8$) exhibit motions with adjacent particles moving in-phase.

ILMs can exist not only in one-dimensional but also in higher-dimensional lattices. Let us show the ILM in a two-dimensional lattice. We consider the two dimensional FPU systems with square lattice [18]. Hamiltonian of the system is given by

$$H = \sum_{i=1}^{N} \sum_{j=1}^{N} \frac{1}{2} m \dot{\mathbf{X}}_{i,j}^2 + \sum_{i=1}^{N} \sum_{j=1}^{N} \left[V(\mathbf{X}_{i+1,j} - \mathbf{X}_{i,j}, d) + V(\mathbf{X}_{i,j+1} - \mathbf{X}_{i,j}, d) \right]$$

$$+ k \sum_{i=1}^{N} \sum_{j=1}^{N} \left[V(\mathbf{X}_{i+1,j+1} - \mathbf{X}_{i,j}, \bar{d}) + V(\mathbf{X}_{i-1,j+1} - \mathbf{X}_{i,j}, \bar{d}) \right], \qquad (4.18)$$

where $\mathbf{X}_{i,j} = (X_{i,j}, Y_{i,j})$ is the position of (i, j)-th particle, m is a mass, d is an equilibrium distance for the nearest neighbor particles, $\bar{d}$ is an equilibrium distance for the second nearest neighbor particles. Interaction potential V is given as follows,

$$V(\mathbf{r}, d) = \frac{1}{2}(|\mathbf{r}| - d)^2 + \frac{1}{4}\beta(|\mathbf{r}| - d)^4, \qquad (4.19)$$

where β is the nonlinear parameter of the system. We set $\beta = 1$ in our calculation. The interaction between the second nearest neighbor particles is $k(<1)$ times smaller than the interaction between the nearest neighbor particles. Periodic boundary conditions are considered in both x and y directions.

Two types of ILMs with different symmetry are found in this system. One type is a quasi-one dimensional ILM shown in Fig. 4.9. Quasi-one dimensional ILM vibrates with large amplitude along one axis. Amplitude along the other axis is negligibly small. In the case of Fig. 4.9, vibration is dominantly observed in x direction.

In the case that $k = 0$, vibration along x direction is dominantly observed only in $j = 0$ line as shown in Fig. 4.9a, b. In the case that $k \neq 0$, on the other hand, vibration along x direction is observed in not only $j = 0$ line but also $j = \pm 1$ lines as shown in Fig. 4.9c, d.

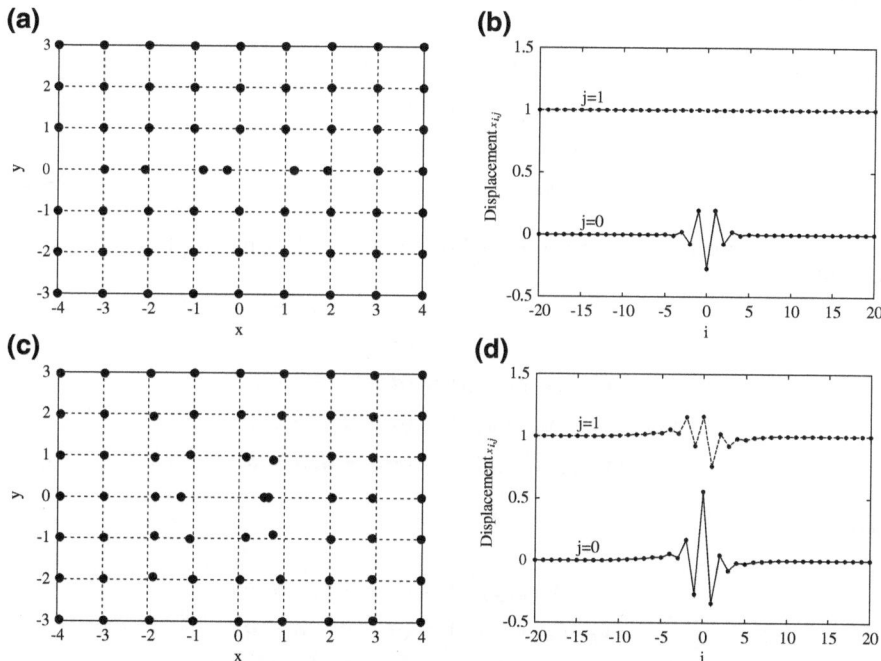

Fig. 4.9 Quasi-one dimensional ILM in the system with (*top*) $k = 0$ and (*bottom*) $k = 1$ and $d = 1$. **a** and **c** show the two dimensional pattern of displacement. **b** and **d** show the displacement of atoms along x axis in $j = 0$ and $j = 1$ lines. Internal frequencies are given as $\omega = 2.4$ (*top*) and $\omega = 4.6$ (*bottom*)

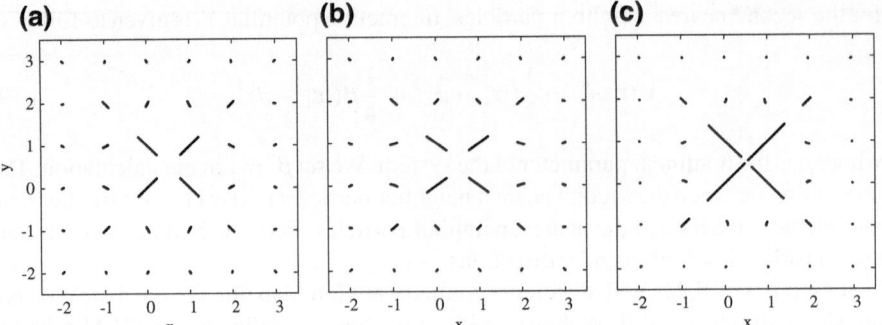

Fig. 4.10 Two dimensional ILM for the system with $k = 1$ and $\bar{d} = \sqrt{2}-0.15$. Angular frequencies are **a** $\omega = 3.4$, **b** $\omega = 3.34$ and **c** $\omega = 4.4$. Displacement patterns correspond to (4.20), (4.21) and (4.22), respectively

The other type is a two-dimensional ILM shown in Fig. 4.10. Two-dimensional ILM has large displacements along both x and y axis. Displacements of the ILM are roughly given as

$$
\begin{aligned}
(x_{m,n}, y_{m,n}) &= (-d, -d), \\
(x_{m+1,n}, y_{m+1,n}) &= (d, -d), \\
(x_{m+1,n+1}, y_{m+1,n+1}) &= (d, d), \\
(x_{m,n+1}, y_{m,n+1}) &= (-d, d),
\end{aligned}
\tag{4.20}
$$

where d is a constant. As a result of bifurcation, following displacement is also possible:

$$
\begin{aligned}
(x_{m,n}, y_{m,n}) &= (-d - a, -d + a), \\
(x_{m+1,n}, y_{m+1,n}) &= (d + a, -d + a), \\
(x_{m+1,n+1}, y_{m+1,n+1}) &= (d + a, d - a), \\
(x_{m,n+1}, y_{m,n+1}) &= (-d - a, d - a),
\end{aligned}
\tag{4.21}
$$

and

$$
\begin{aligned}
(x_{m,n}, y_{m,n}) &= (-d - b, -d - b), \\
(x_{m+1,n}, y_{m+1,n}) &= (d - b, -d + b), \\
(x_{m+1,n+1}, y_{m+1,n+1}) &= (d + b, d + b), \\
(x_{m,n+1}, y_{m,n+1}) &= (-d + b, d - b).
\end{aligned}
\tag{4.22}
$$

where a and b are constants. The three types of displacement (4.20), (4.21) and (4.22) correspond to Fig. 4.10a–c, respectively.

In higher dimensional lattices, it is expected that a large variety of ILMs with different symmetries are possible, depending on the structure of lattice (triangle, rectangle, and hexagonal...) and interaction potentials.

4.3 Basic Properties of Fundamental Modes

4.3.1 Approximate Solution of Weakly Localized ILM

We show how the profile of ILM depends on its parameters such as frequency or amplitude, using an approximate analytical solution of ILM. The FPU lattice (4.4) is used as an example, and its equations of motion are given by

$$\ddot{q}_n = q_{n+1} - 2q_n + q_{n-1} + (q_{n+1} - q_n)^3 - (q_n - q_{n-1})^3. \qquad (4.23)$$

We construct an approximate ILM solution in weak localization regime, according to a continuous approximation approach developed in [19, 20].

Numerical solutions in Fig. 4.3c, d shows that ILMs have envelopes slowly varying in space in which staggered displacement patterns are realized, for Ω close to the band-edge frequency ω_{max}. We make the transformation $u_n = (-1)^n q_n$ to remove the staggered displacements. Then, we introduce a smooth envelope function of the displacements $f(x, t)$ such that $f(n, t) = u_n(t)$, $n \in \mathbb{Z}$. This function is assumed to be slowly varying on the inter-particle scale. To express this property, we assume $\partial^j f / \partial x^j = O(\varepsilon^j)$, $j = 1, 2, \ldots$, where $\varepsilon \ll 1$ is a small parameter. If we introduce a function F which is defined by $F(\xi, t) = f(x, t)$ with $\xi = \varepsilon x$, this assumption is equivalent to $\partial^j F / \partial \xi^j = O(1)$, $j = 1, 2, \ldots$. We can expand $u_{n\pm1}$ as

$$u_{n\pm1} = F(\xi, t) \pm \varepsilon F_\xi(\xi, t) + \frac{1}{2}\varepsilon^2 F_{\xi\xi}(\xi, t) + \cdots, \qquad (4.24)$$

where $\xi = \varepsilon n$ and the subscript ξ stands for the spatial derivative of the corresponding order. If we substitute these expansions with $u_n = (-1)^n q_n$ into (4.23), we can obtain a nonlinear partial differential equation for F up to the order of ε^2 as follows:

$$F_{tt} + \varepsilon^2 F_{\xi\xi} + 6\varepsilon^2 F(F^2)_{\xi\xi} + 4F + 16F^3 = 0, \qquad (4.25)$$

where the subscript t stands for the time derivative. In terms of x and f, this equation can be rewritten into the equation

$$f_{tt} + f_{xx} + 6f(f^2)_{xx} + 4f + 16f^3 = 0. \qquad (4.26)$$

We look for a localized solution in the form $f(x, t) = \phi(x) \cos \Omega t$. If we substitute this ansatz into (4.26) and invoking the rotating wave approximation, which replaces

$\cos^3 \Omega t$ with $\frac{3}{4}\cos\Omega t$ and neglect the harmonics other than Ω, then we obtain the first integral

$$\left(4 - \Omega^2\right)\phi^2 + \left(1 + 9\phi^2\right)\phi_x^2 + 6\phi^4 = C, \qquad (4.27)$$

where C is an integration constant. We choose $C = 0$ since we are looking for a solution such that $\lim_{x\to\pm\infty}\phi(x) = 0$ and $\lim_{x\to\pm\infty}\phi_x(x) = 0$. We impose an additional condition $\phi(x) > 0$ to the solution. Let $A = \max_x \phi(x)$ and x_0 be a point such that $\phi(x_0) = A$. Using $\phi_x(x_0) = 0$ and $\phi(x_0) = A$ in (4.27), we can obtain the frequency as a function of the amplitude A:

$$\Omega^2 = 4 + 6A^2. \qquad (4.28)$$

Using (4.28) in (4.27), we have

$$\left(\frac{d\phi}{dx}\right)^2 = \frac{6\phi^2\left(A^2 - \phi^2\right)}{1 + 9\phi^2}, \qquad (4.29)$$

This equation can be solved in the following implicit form for $\phi(x)$:

$$|x - x_0| = \frac{1}{\sqrt{6}}\int_\phi^A \frac{\sqrt{1 + 9\phi'^2}}{\sqrt{\phi'^2\left(A^2 - \phi'^2\right)}}\,d\phi'. \qquad (4.30)$$

Let us consider a small-amplitude regime $A \ll 1$. In this regime, $9\phi'^2$ in the numerator can be neglected, compared with the unity. Then, we can calculate the integral in (4.30) to obtain $\phi(x)$, and arrive at.

$$f(x,t) = A\,\mathrm{sech}\left[\sqrt{6}A(x - x_0)\right]\cos\Omega t. \qquad (4.31)$$

Equation (4.31) shows that the spatial profile of f becomes narrower as its amplitude A increases, since the width of f is in proportion to A^{-1}. On the other hand, (4.28) is rewritten as $\Omega^2 - \omega_{max}^2 = 6A^2$, where $\omega_{max} = 2$ is the maximum frequency of the phonon band, and it shows that Ω deviates from the band edge ω_{max} with increasing A. Therefore, it can be concluded that the profile of ILM becomes narrower as the frequency more deviates from the phonon band edge. ILMs are weakly localized and thus solution (4.31) is expected to be accurate when Ω is close to ω_{max}. This property of ILM profile is consistent with the fact $\lim_{\omega\to\infty} r = 0$, which follows from (4.15). We emphasize that this is a general property of ILMs, which is observed also in the other lattices. For instance, an ILM of the NKG lattice with Ω in the lower forbidden band becomes narrower as Ω decreases and more deviates from the lower phonon band edge.

We remark that the approximate solution (4.31) is useful from the point of view of numerical computation. The Newton method is often used for numerically computing a periodic solution of ILM. The method requires a precise approximation as an initial guess. Equation (4.31) gives a good approximate solution for this purpose: given Ω, it gives the initial conditions for an approximate ILM solution as

$$q_n(0) = (-1)^n A \operatorname{sech}\left[\sqrt{6}A(n - n_0)\right], \quad p_n(0) = 0, \tag{4.32}$$

where $A = \sqrt{(\Omega^2 - \omega_{max}^2)/6}$ and the cases $n_0 \in \mathbb{Z}$ and $n_0 + 1/2 \in \mathbb{Z}$ correspond to the ST and P mode solutions, respectively.

4.3.2 Approximate Solution of Strongly Localized ILM

We construct approximate ILM solutions for the ST and P modes in strong localization regime. Let us consider a one-dimensional lattice slightly generalized from the FPU lattice (4.4), which is described by the Hamiltonian

$$H = \sum_n \frac{1}{2}p_n^2 + \sum_n \sum_{r=2}^{k} \frac{\kappa_r}{r}(q_{n+1} - q_n)^r, \tag{4.33}$$

where $k \geq 4$ is an even integer and κ_r are constants. We can assume $\kappa_k = 1$ by normalizing the variables. The FPU lattice (4.4) corresponds to the case of $k = 4$, $\kappa_2 = 1$ and $\kappa_3 = 0$. The equations of motion for Hamiltonian (4.33) are given by

$$\ddot{q}_n = \sum_{r=2}^{k} \kappa_r \left[(q_{n+1} - q_n)^{r-1} - (q_n - q_{n-1})^{r-1} \right]. \tag{4.34}$$

Let us look for an ILM solution with a large amplitude. For such a solution, the highest order terms on the right hand side become dominant and the other terms may be negligible in (4.34). If we retain only the highest order terms, we have

$$\ddot{q}_n = (q_{n+1} - q_n)^{k-1} - (q_n - q_{n-1})^{k-1}. \tag{4.35}$$

It is possible to seek a periodic solution of (4.35) in the form

$$q_n(t) = u_n \phi(t), \tag{4.36}$$

where $u_n \in \mathbb{R}$ are time-independent constants and $\phi(t) \in \mathbb{R}$ is a function of time t. If we substitute (4.36) into (4.35), we obtain the differential equation

$$\ddot{\phi} + \phi^{k-1} = 0 \tag{4.37}$$

for $\phi(t)$ and the set of algebraic equations

$$u_n + (u_{n+1} - u_n)^{k-1} - (u_n - u_{n-1})^{k-1} = 0. \tag{4.38}$$

If a periodic solution ϕ of (4.37) and a solution $u = \{u_n\}$, $n \in \mathbb{Z}$ of (4.38) are obtained, then they give a periodic solution of the original equation (4.35) being of the form (4.36).

Let $\phi(t)$ be a solution of (4.37) with the initial conditions $\phi(0) = A > 0$ and $\dot{\phi}(0) = 0$. Equation (4.37) has the energy integral $\dot{\phi}^2/2 + \phi^k/k = h$, where $h > 0$ is an integration constant, and it is regarded as a Hamiltonian system with the potential ϕ^k/k. It is clear that $\phi(t)$ is a non-constant periodic solution for any A. The period T of $\phi(t)$ depends on h ($=A^k/k$), and it is obtained from the energy integral as follows:

$$T = 2\sqrt{2}\, h^{-(1/2-1/k)} \int_0^{k^{1/k}} \frac{1}{\sqrt{1 - x^k/k}}\, dx. \tag{4.39}$$

The frequency Ω is given by $\Omega = 2\pi/T$. If we use $h = A^k/k$ and note that the integral in (4.39) is independent of h, we have the relation

$$\Omega \propto A^{k/2-1}. \tag{4.40}$$

This indicates that Ω continuously varies from $\Omega = 0$ to $+\infty$ as A varies from $A = 0$ to $+\infty$. We denote the periodic solution of (4.37) with frequency Ω with $\phi(t; \Omega)$.

We proceed to calculate approximate solutions of (4.38), distinguishing two cases of the ST and P modes. First, to consider the ST mode case, assume the solution symmetry $u_{-n} = u_n$, $n \in \mathbb{N}$. In addition, since a strongly localized solution is being sought, it may be reasonable to assume $u_{\pm n} = 0$ for $n \geq 2$. If we use these assumptions in (4.38), we have the reduced set of equations

$$u_0 + 2(u_1 - u_0)^{k-1} = 0, \tag{4.41}$$
$$u_1 - u_1^{k-1} - (u_1 - u_0)^{k-1} = 0. \tag{4.42}$$

These equations can be explicitly solved, and we have an approximate solution of (4.38) as follows:

$$u_n = \begin{cases} 2 \times 3^{-(k-1)/(k-2)} & \text{if } n = 0, \\ -3^{-(k-1)/(k-2)} & \text{if } n = \pm 1, \\ 0 & \text{otherwise.} \end{cases} \tag{4.43}$$

It is easy to check that this gives a good approximate solution to (4.38). An approximate ST mode solution to (4.35) is obtained by combining this $\{u_n\}$ and $\phi(t; \Omega)$ according to (4.36). This ST mode solution $u_n\phi(t; \Omega)$ gives a good approximate solution of (4.34), provided that its amplitude is large or, equivalently, the frequency Ω

is large as shown by (4.40). The normalized spatial profile u_n/u_0 of this approximate
ST mode solution is given by $(\ldots, 0, -1/2, 1, -1/2, 0, \ldots)$.

Next, let us consider the P mode case. Assume the solution symmetry $u_{-n+1} = -u_n$, $n \in \mathbb{N}$ and a strongly localized profile such that $u_n = 0$ for $n \neq 0, 1$. Using
these assumptions in (4.38), we have the reduced equation

$$u_0 - \left(1 + 2^{k-1}\right) u_0^{k-1} = 0, \tag{4.44}$$

The nonzero solution of (4.44) yields an approximate solution of (4.38) as follows:

$$u_n = \begin{cases} (1 + 2^{k-1})^{-1/(k-2)} & \text{if } n = 0, \\ -(1 + 2^{k-1})^{-1/(k-2)} & \text{if } n = 1, \\ 0 & \text{otherwise.} \end{cases} \tag{4.45}$$

It is easy to check that this gives a good approximate solution to (4.38). The P mode
solution $u_n \phi(t; \Omega)$ obtained from (4.45) gives a good approximate solution of (4.34),
provided that its amplitude or the frequency Ω is large. The normalized spatial profile
u_n/u_0 of this approximate P mode solution is given by $(\ldots, 0, 1, -1, 0, \ldots)$.

The approximate solutions constructed above show that a strongly localized ILM
is realized for large Ω. This is consistent with the numerical results in Fig. 4.3.

4.3.3 Stability of ILM

Once an ILM solution is found, a fundamental issue is its stability. In principle, it is
necessary to study the evolution of perturbations added to the ILM solution, based
on the equations of motion. However, for sufficiently small perturbations, one may
use the linearized equations of motion to study the stability.

Let us consider Hamiltonian (4.1) and its equations of motion, (4.2), in the case
of a finite dimension, i.e., $n = 1, \ldots, N$. Let $\{q_n(t), p_n(t)\}$ be an ILM solution with
a period T. Linearizing (4.2) along this solution, we have the variational equations

$$\dot{\xi}_i = \sum_{j=1}^{N} \left[\frac{\partial^2 H}{\partial p_i \partial q_j} \xi_j + \frac{\partial^2 H}{\partial p_i \partial p_j} \eta_j \right], \quad \dot{\eta}_i = -\sum_{j=1}^{N} \left[\frac{\partial^2 H}{\partial q_i \partial q_j} \xi_j + \frac{\partial^2 H}{\partial q_i \partial p_j} \eta_j \right], \tag{4.46}$$

where ξ_i and η_i are variations in variables q_i and p_i, respectively. Introduce the
notations $z = (q_1, \ldots, q_N, p_1, \ldots, p_N) \in \mathbb{R}^{2N}$ and $\zeta = (\xi_1, \ldots, \xi_N, \eta_1, \ldots, \eta_N) \in \mathbb{R}^{2N}$. Equation (4.46) can be written in the vector form

$$\dot{\zeta} = J H_{zz}(z(t)) \cdot \zeta, \tag{4.47}$$

where $H_{zz}(z(t))$ is the Hessian matrix of the Hamiltonian H evaluated at $z(t)$ and J is the $2N \times 2N$ symplectic matrix

$$J = \begin{pmatrix} 0 & I \\ -I & 0 \end{pmatrix}.$$ (4.48)

where I is the $N \times N$ identity matrix.

Let $\{\zeta^{(1)}, \ldots, \zeta^{(2N)}\}$ be a system of fundamental solutions of (4.47), where $\zeta^{(j)} \in \mathbb{R}^{2N}$ for each j. Since the ILM solution is T-periodic, i.e., $z(t + T) = z(t)$, so is the matrix $JH_{zz}(z(t))$ in (4.47). According to the Floquet theory, the fundamental solutions of (4.47) at t and $t + T$ are related via a $2N \times 2N$ monodromy matrix M as

$$\left(\zeta^{(1)}(t+T), \ldots, \zeta^{(2N)}(t+T) \right) = \left(\zeta^{(1)}(t), \ldots, \zeta^{(2N)}(t) \right) \cdot M.$$ (4.49)

Eigenvalues of M are called the characteristic multipliers, and they are independent of the choice of fundamental solutions. It follows from Hamiltonian nature of the system that $+1$ is an eigenvalue of M with multiplicity at least two, and that if ρ is an eigenvalue of M, so is ρ^{-1}.

Spectral stability is a useful and commonly used notion of stability, which is defined by using the characteristic multipliers [21]. Let ρ_j, $j = 1, \ldots, 2N$ be the characteristic multipliers of $z(t)$. The spectral stability is defined as follows.

Definition. *Periodic solution $z(t)$ is said to be spectrally unstable if there exists ρ_j such that $|\rho_j| > 1$, otherwise it is said to be spectrally stable.*

Figure 4.11a, b shows the characteristic multipliers of ST and P modes of FPU lattice (4.4) on the complex plane, respectively, where the frequency is $\Omega = 10$. The corresponding mode profiles are shown in Fig. 4.3a, b. In Fig. 4.11a, there is one characteristic multiplier which is located on the real axis and has the modulus larger than unity. This indicates that the ST mode is spectrally unstable. In contrast, all the characteristic multipliers are located on the unit circle in Fig. 4.11b, indicating that the P mode is spectrally stable. These properties of stability of the ST and P modes in the FPU lattice were first shown in [22].

It is known by numerical simulations that the instability of ST mode is not destructive to its localized structure. Typically, generic perturbations, which include an eigenvector of M for the unstable characteristic multiplier, result in the ILM motion along the lattice. However, during this motion the ILM gradually loses its energy through radiation of small ripples, eventually it becomes trapped, and perform oscillations around the stable P mode configuration. On the other hand, generic perturbations to the P mode cause oscillations of the ILM center around its stable position.

The site-centered (ST) mode is spectrally unstable while the bond-centered (P) mode is spectrally stable in the above case. Unfortunately, this observation law does not necessarily apply to the other lattice models, and there is no general stability law. We show that the stability of both modes depends on the model. Consider the lattice

Fig. 4.11 Characteristic multipliers are plotted for **a** ST mode with $\Omega = 10$ and **b** P mode with $\Omega = 10$ in FPU lattice (4.4). The lattice size is $N = 51$, and fixed-end boundary conditions are used

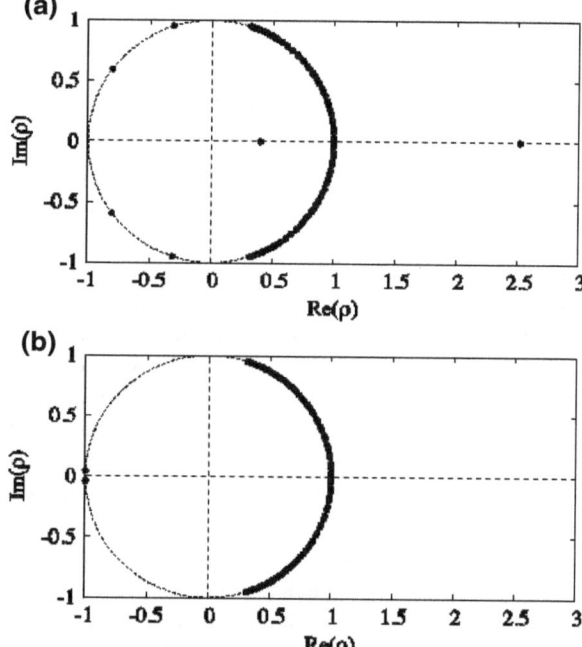

Hamiltonian

$$H = \sum_n \frac{1}{2}p_n^2 + \sum_n \left[\frac{1}{2}q_n^2 + \frac{1}{4}q_n^4 + \frac{\kappa_2}{2}(q_{n+1} - q_n)^2 + \frac{\kappa_4}{4}(q_{n+1} - q_n)^4 \right],$$
(4.50)

where κ_2 and κ_4 are constants. This model has both site-centered and bond-centered mode solutions. Their stability is numerically examined by varying the quartic interaction coefficient κ_4 while fixing $\kappa_2 = 0.1$ in [23]. An exchange of stability between these two types of modes is reported. The site-centered mode is stable and the bond-centered one is unstable for small $\kappa_4 \geq 0$. The former becomes unstable at $\kappa_4 \simeq 0.553048$, and then the latter becomes stable at $\kappa_4 \simeq 0.553138$, as κ_4 increases. Both modes are unstable in the small interval $\mathcal{I} \simeq (0.553048, 0.553138)$. Furthermore, it is shown that the stability exchange process is accompanied by bifurcations of site-centered and bond-centered ILM solutions, and that an intermediate type ILM solution having no spatial symmetry appears for $\kappa_4 \in \mathcal{I}$.

4.3.4 Rigorous Results on ILM in FPU Lattice

This subsection is devoted to a brief overview of mathematically rigorous results on existence proof and stability analysis for ILMs in one-dimensional FPU lattices described by the Hamiltonian of the form

$$H = \sum_n \frac{1}{2} p_n^2 + \sum_n V(q_{n+1} - q_n). \tag{4.51}$$

For this class of lattices, the first existence proof was given in the particular case of homogeneous interaction potentials [24], where $V(X) = X^k/k$ with an even integer $k \geq 4$ and the equations of motion are given by (4.35). This particular choice of potential allows for a separation of time and space variables (cf. Sect. 4.3.2). The problem of proving the existence of an ILM solution in an infinite lattice is reduced to that of showing a homoclinic orbit in a two-dimensional map. Then, the existence of ILM solutions of the ST and P mode symmetries has been proved in the case of infinite lattice, based on this homoclinic orbit approach.

As for the case of non-homogeneous interaction potentials, an implicit proof of existence of ILMs was given in [25] using a variational method. The existence of a time-periodic solution satisfying the localization condition (4.6) has been proved for a class of lattices including (4.51) with convex interaction potentials. However, this proof does not provide a detail information about the spatial profile of the time-periodic solution. It is unclear whether the solution has the symmetry of ST or P mode. Only a numerical calculation suggested that it has the P mode profile.

James performed a center manifold reduction and proved the existence of two types of weakly localized ILMs in infinite FPU lattices, which have small amplitudes and frequencies close to the phonon band edge, provided that $V(0) = V'(0) = 0$, $V''(0) > 0$, and $V''(0)V^{(4)}(0)/2 - \{V^{(3)}(0)\}^2 > 0$ [26]. It was also proved that each of these ILMs has the symmetry of either ST or P mode in its spatial profile. Therefore, this is an explicit proof for the existence of ST and P modes in infinite FPU lattices in weak localization regime.

Recently, the existence of ST and P modes in strongly localization regime has been proved in finite FPU lattices [27]. The case of nearly homogeneous potential was assumed: $V(X) = W(X, \mu) + X^k/k$ with an even integer $k \geq 4$ and a smooth perturbation $W(X, \mu)$ such that $W(X, 0) = 0$. ILM solutions are constructed in the homogeneous potential FPU lattice ($\mu = 0$) by using Banach's fixed point theorem, and then they are continued to a non-homogeneous potential one ($\mu \neq 0$). This result ensures the existence of large-amplitude ST and P modes in FPU lattices when $V(X) = \sum_{r=2}^k \kappa_r X^r$ with an even integer $k \geq 4$ and $\kappa_k > 0$. Moreover, it has been proved that the ST mode is spectrally unstable while the P mode is spectrally stable.

4.3.5 ILM in Dissipative Lattices

We have been discussing ILMs in Hamiltonian lattices so far. However, any experiments will necessarily be accompanied with some dissipation such as friction. In dissipative lattices, ILMs can still exist as periodic solutions, provided that some sources of incoming energy, e.g., an external periodic forcing, are present and they are balanced with the dissipation. Compared with Hamiltonian case, a fundamental difference is that ILMs no longer form continuous families of periodic solutions, but an ILM may exist as an isolated periodic solution, typically being an attractor.

Let us show an example of ILM in a dissipative lattice, which is a nonlinear Klein–Gordon type lattice with a dissipative and an external forcing:

$$\ddot{q}_n = -q_n - q_n^3 - \kappa_2(2q_n - q_{n-1} - q_{n+1}) - \gamma \dot{q}_n + F_{\text{ext}} \cos(\omega t), \qquad (4.52)$$

where γ is a coefficient of the damping term and F_{ext} is an amplitude of the periodic forcing of angular frequency ω. Figure 4.12a–f shows wave form of ILMs having period $T = 3$. Parameters are set at $\kappa_2 = 0.1$, $\gamma = 0.1$, $F_{\text{ext}} = 1$, and $\omega = 2\pi/3$. In the dissipative lattice, as shown in Fig. 4.12a, b, there exist two site-centered modes in which phase to the external force is different. In a similar fashion, three types of bond-centered mode are obtained as shown in Fig. 4.12d–f. The variety of ILM comes from the fact that three periodic solutions appear near the resonant frequency in single oscillator with periodic forcing. One is the stable resonance("s") which shows large amplitude oscillation and is stable. Another is the unstable resonance("u"). Although it also has large amplitude, the oscillation is unstable. The remaining one is the nonresonance("o") which shows small amplitude oscillation. This state is stable. If there is no coupling, namely $\kappa_2 = 0$, any periodic solutions consisting of sequences of characters "s", "u", and "o" are trivially exist (cf. Sect. 4.4.1). We can obtain ILM solution by continuing the trivial solutions into the weak coupling regime. For example, the site-centered mode shown in Fig. 4.12a is obtained from

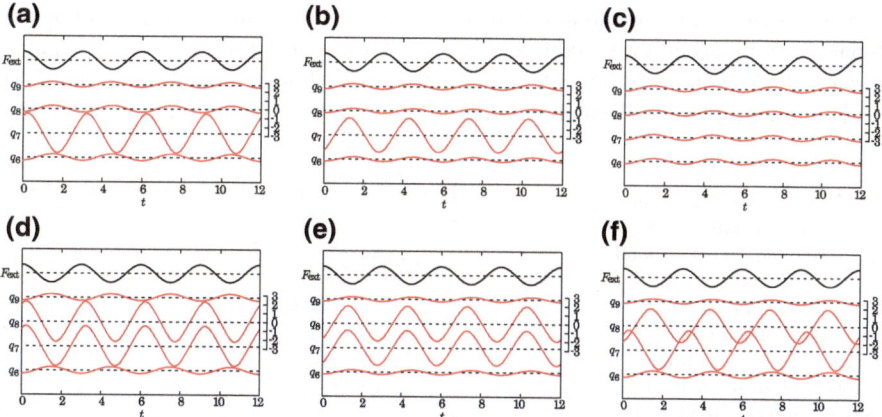

Fig. 4.12 Wave form of ILMs in (4.52). Each ILM is obtained by a code **a** (..., o, s, o, o, ...), **b** (..., o, u, o, o, ...), **c** (..., o, o, o, o, ...), **d** (..., o, s, s, o, ...), **e** (..., o, u, u, o, ...), **f** (..., o, s, u, o, ...), **a** stable site-centered mode, **b** unstable site-centered mode, **c** ground state, **d** stable bond-centered mode, **e** unstable bond centered mode, **f** unstable bond centered mode

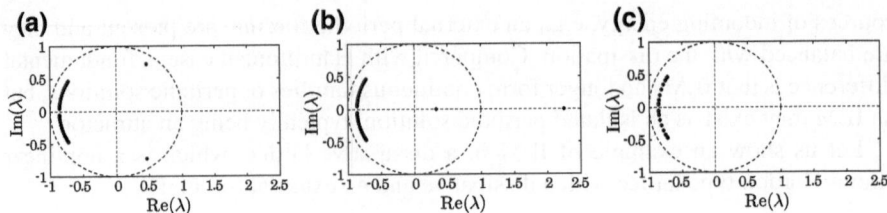

Fig. 4.13 **a** stable ILM **b** unstable ILM **c** ground state. Characteristic multipliers of ILM shown in Fig. 4.12a–c

(..., o, s, o, o, ...), and the bond-centered mode shown in Fig. 4.12f from (. . . , o, s, u, o, . . .).

Stability of ILM inherits that of periodic solutions of single oscillator in the weak coupling regime, i.e., a site-centered mode is stable if it is originated from the code (..., o, s, o, o, ...) whereas the other site-centered mode is unstable if it is from (..., o, u, o, o, ...). Figure 4.13a–c shows characteristic multipliers of the stable, unstable site-centered modes, and the ground state. All the characteristic multipliers are inside the unit circle for the stable site-centered mode and the ground state, while one of characteristic multipliers is outside the unit circle for the unstable site-centered mode. In the former case, it can be said *stable* in rigorous sense. That is, when the solutions is perturbed slightly, the trajectories will converge to the original solutions. Therefore, the stable solution has an area called *basin of attraction* (see Fig. 4.14). All the trajectories started from the area finally converge to the stable solution. This is the significant difference between the dissipative lattice and the Hamiltonian lattice.

Since a dissipative oscillator with a periodic forcing shows very complicated dynamics comparing with a conservative oscillator with no forcing, the dissipative lattice will show quite complicated phenomena even though we only focus on ILM. In other wards, ILM in dissipative lattices still remains as fertile area of research.

Fig. 4.14 Example of basin of attraction of stable ILM "so" (*dark region*) and ground state "oo" (*white region*). Three points (*open circle, open square, filled circle*) indicate periodic solutions (stable ILM, unstable ILM, ground state) at $t \mod T = 0$

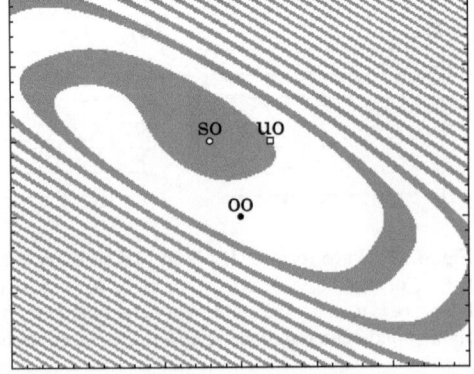

4.4 Variations of ILM

4.4.1 ILMs Near the Anti-integrable Limit

The concept of *anti-integrable* (AI) limit was originally introduced by Aubry and Abramovici to study chaotic trajectories of the standard map [28]. This concept has been applied to study properties of ILMs and turned out to be useful. We explain the AI limit and discuss the properties of ILMs near this limit.

Let us consider nonlinear Klein–Gordon type lattices with weak interactions, which is defined by the Hamiltonian

$$H = \sum_{n=1}^{N} \left[\frac{1}{2} p_n^2 + U(q_n) \right] + \varepsilon \sum_{n=1}^{N-1} V(q_{n+1} - q_n), \qquad (4.53)$$

where ε is a small parameter, V is a smooth function, U is a smooth anharmonic function such that $U(0) = U'(0) = 0$ and $U''(0) = \omega_0^2 > 0$. The first sum in (4.53) represents a set of Hamiltonian oscillators, and the second sum represents weak nearest neighbour interactions. Hamiltonian (4.53) describes finite-size lattices with the open-end boundary conditions. The equations of motion read

$$\ddot{q}_n = -U'(q_n) + \varepsilon \left[V'(q_{n+1} - q_n) - V'(q_n - q_{n-1}) \right], \quad n = 1, \dots, N, \quad (4.54)$$

where the equations for $n = 1$, N do not have the corresponding interaction terms due to the open-end boundary conditions.

The AI limit of this lattice is defined by the limit $\varepsilon = 0$, and it is also called the *anti-continuous* limit. When $\varepsilon = 0$, (4.54) decouples with each other, and each equation has the form

$$\ddot{\phi} + U'(\phi) = 0. \qquad (4.55)$$

This is a Hamiltonian dynamical system with the energy integral $\dot{\phi}^2/2 + U(\phi) = E$. The level sets of E near the origin $(\phi, \dot{\phi}) = (0, 0)$ form closed curves around the origin, because of the assumption $U''(0) > 0$. Therefore, (4.55) has a family $\phi(t; E)$ of periodic solutions parametrized by E within a certain range $E \in (0, E_0)$, for which $\phi(t; E) = \phi(-t; E)$ can be assumed. We denote the period of $\phi(t; E)$ with $\tau(E)$.

In the AI limit, given $E \in (0, E_0)$, (4.54) has a number of trivial periodic solutions, which we call *anti-integrable solutions*, of the form

$$q_n(t) = \sigma_n \phi(t; E), \quad n = 1, \dots, N \qquad (4.56)$$

with the period $\tau(E)$, where $\sigma_n \in \{0, \pm 1\}$. Any anti-integrable solution is represented by an integer sequence $\sigma \equiv (\sigma_1, \dots, \sigma_N)$ called a *coding sequence*. An anti-integrable solution is well localized when its σ consists of a small number of nonzero components: for example, $\sigma = (\ldots, 0, 1, 0 \ldots)$ represents a single-site ILM and $\sigma = (\ldots, 0, 1, 0, 0, -1, 0, \ldots)$ represents two single-site ILMs located separately.

Assume that solution (4.56) has period T and σ is arbitrary. MacKay and Aubry proved by using the implicit function theorem that for small $\varepsilon > 0$ (4.54) has a family of T-periodic solutions which is a continuation of solution (4.56) with respect to ε, under the following assumptions [29]: (H1) $\phi(t; E)$ is *non-resonant*, i.e., $2\pi m/T \neq \omega_0$ for $^\forall m \in \mathbb{N}$; (H2) $\phi(t; E)$ is *anharmonic*, i.e., $d\tau/dE|_{\tau=T} \neq 0$. We remark that their proof also applies to infinite-size lattices. Their result is important since it reveals that a variety of periodic solutions exist near the AI limit. Some of them can be regarded as localized periodic solutions, i.e., ILMs. In addition, their approach based on the AI limit gives an efficient method of numerically computing ILMs [16]. Examples of numerical ILM solutions continued from the AI limit are shown for some coding sequences σ in Fig. 4.15, where $U(X) = X^2/2 + X^4/4$ and $V(X) = X^2/2$ are employed.

There are a large number of single-site or multi-site ILMs near the AI limit, which have configurations characterized by different coding sequences. An fundamental issue is their stability. It has been rigorously revealed that for small $\varepsilon > 0$ there

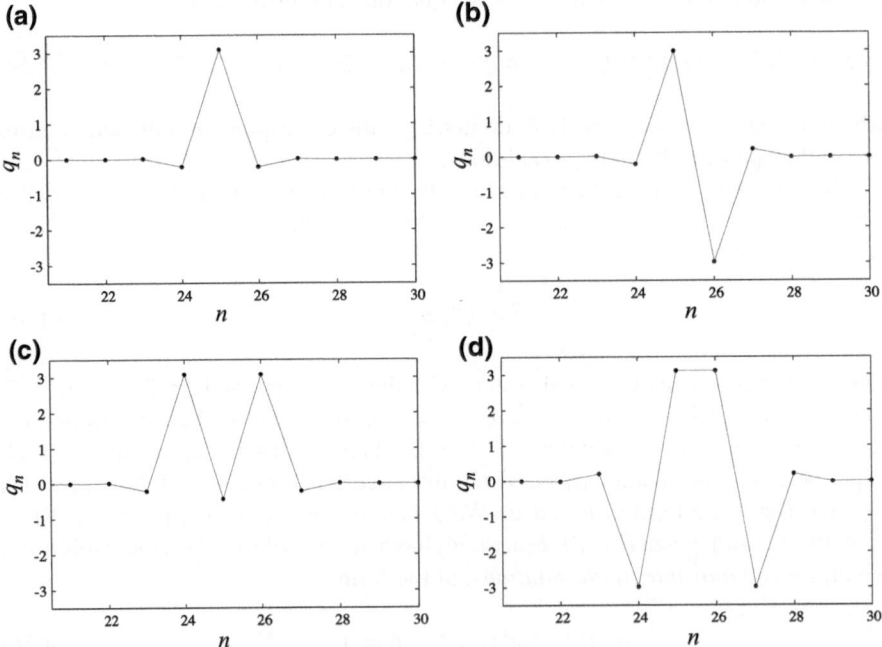

Fig. 4.15 Numerical solution of ILM near the AI limit in nonlinear Klein–Gordon type lattice (4.53). Parameters are $N = 51$, $\varepsilon = 0.5$, and $\Omega = 3$. Displacement $q_n(0)$ is plotted against site number n for $\sigma = (\ldots, 0, 1, 0, \ldots)$ (**a**), $(\ldots, 0, 1, -1, 0, \ldots)$ (**b**), $(\ldots, 0, 1, 0, 1, 0, \ldots)$ (**c**), and $(\ldots, 0, -1, 1, 1, -1, 0, \ldots)$ (**d**)

Fig. 4.16 Illustration of two-site ILM in the AI limit

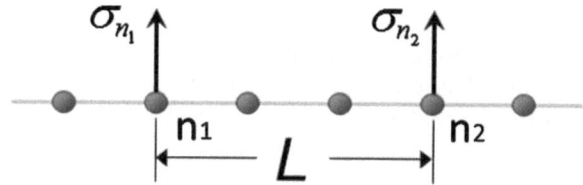

underlies a simple law which relates the spectral stability of an ILM with its coding sequence σ, in each case of harmonic interaction $V(X) = X^2/2$ [30–32] and pure anharmonic interaction $V(X) = X^k/k$ with an even integer $k \geq 4$ [33].

Let us consider the case of two-site ILM to describe the stability law in a simple case. Figure 4.16 illustrates a two-site ILM in the AI limit, where $L = n_2 - n_1$ represents the distance between two adjacent excited sites. The stability or instability of the two-site ILM for small $\varepsilon > 0$ depends on the phase difference, which is represented by the sign of $\sigma_{n_1}\sigma_{n_2}$, and the parity of distance L. The phase difference is either in-phase $\sigma_{n_1}\sigma_{n_2} > 0$ or anti-phase $\sigma_{n_1}\sigma_{n_2} < 0$.

In the case of harmonic interaction $V(X) = X^2/2$, the most general result is obtained in [32]. Table 4.1 summarizes the result, provided that $U(X) = X^2/2 + \alpha X^4/4$, where $\alpha = \pm 1$ represents hard and soft nonlinearities, respectively. The types of spectrally stable two-site ILMs are shown in Table 4.2: for instance, in-phase ILMs are spectrally stable when $\alpha = +1$ and L odd. The result for $\alpha = -1$ and L even means that there exists a critical period T^* between two resonances 2π and 6π such that anti-phase modes are stable for a period $T \in (2\pi, T^*)$ while so are in-phase modes for $T \in (T^*, 6\pi)$. The critical periods exist between every two

Table 4.1 Type of spectrally stable two-site ILM for small $\varepsilon > 0$ in lattice (4.53). Potentials are $U(X) = X^2/2 + \alpha X^4/4$ and $V(X) = X^2/2$

	L odd	L even
Hard potential $\alpha = +1$	In-phase	Anti-phase
Soft potential $\alpha = -1$	Anti-phase	Anti-phase ($2\pi < T < T^*$) In-phase ($T^* < T < 6\pi$)

Table 4.2 Type of spectrally stable two-site ILM for small $\varepsilon > 0$ in lattice (4.53). Potentials are $U(X) = X^2/2 + \alpha X^k/k$ and $V(X) = X^k/k$, where $k \geq 4$ is an even integer

	L odd	L even
Hard potential $\alpha = +1$	In-phase	Anti-phase
Soft potential $\alpha = -1$	Anti-phase In-phase[†]	Anti-phase

resonances $2\pi(2l-1)$ and $2\pi(2l+1)$, where $l \in \mathbb{N}$, and a similar stability change is observed for each interval.

Table 4.2 is a similar table for the lattice with anharmonic interactions, which has $U(X) = X^2/2 + \alpha X^k/k$, $\alpha = \pm 1$ and $V(X) = X^k/k$ with $k \geq 4$ an even integer. The results are identical with those of Table 4.1 in the hard potential case. On the other hand, a difference appears in the soft potential case: no stability change is observed for L even. We remark that in the column of L odd the symbol † stands for a degenerate case due to an additional symmetry of σ, and thus appearance of stable in-phase ILMs is exceptional.

The concept of AI limit is not limited to the nonlinear Klein–Gordon type lattices. It can be defined in some other lattices. An interesting example is the diatomic FPU lattice which consists of alternating light and heavy particles. Its Hamiltonian is given by (4.1) with $U(X) = 0$ and $m_{2j-1} = m$, $m_{2j} = M$ for $j \in \mathbb{Z}$. Let $\varepsilon = \sqrt{m/M}$. Livi et al. introduced the AI limit $\varepsilon = 0$, and have proved the existence of ILMs near this limit [34]. In addition, the stability law of ILMs has been proved, based on two-step continuation of anti-integrable solutions using the associated homogeneous potential lattice [35].

4.4.2 Polarobreather

Let us consider a coupled electron–phonon system in a one-dimensional lattice, which is described in dimensionless form by the equations of motion

$$i\gamma \dot{\psi}_n + J(\psi_{n+1} + \psi_{n-1}) - u_n \psi_n = 0, \tag{4.57}$$

$$\ddot{u}_n + U'(u_n) + |\psi_n|^2 = 0, \tag{4.58}$$

where $\psi_n(t) \in \mathbb{C}$ is the amplitude for finding the electron in the localized state at site n, $u_n(t) \in \mathbb{R}$ represents the displacement of oscillator at site n, U is an anharmonic on-site potential, and J and γ are parameters corresponding to the transfer integral and the time scale ratio between the lattice dynamics and the electron dynamics, respectively. The normalized condition $\sum_n |\psi_n(t)|^2 = 1$ is imposed. The above model is illustrated in Fig. 4.17.

Consider the differential equation

$$\ddot{u} + U'(u) + \rho = 0, \tag{4.59}$$

where $u \in \mathbb{R}$ and $0 \leq \rho \leq 1$. This is regarded as a Hamiltonian oscillator with the potential $U_\rho(u) = U(u) + \rho u$. We assume that for each $\rho \in [0, 1]$ there is a family $\chi_\rho(t; E)$ of periodic solutions parametrized by the energy E such that $\chi_\rho(t; E) = \chi_\rho(-t; E)$. Denote the period of $\chi_\rho(t; E)$ with $\tau_\rho(E)$.

A *polarobreather* (PB) is defined as a spatially-localized and time-periodic solution of (4.57) and (4.58). PBs in these equations have been discussed in [8, 36].

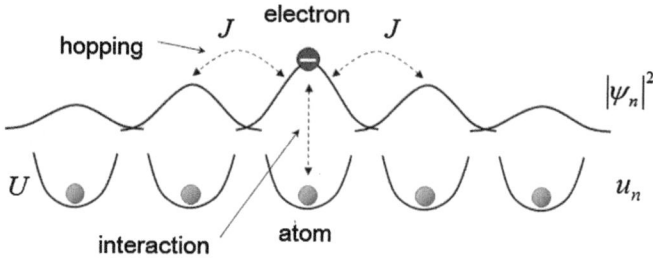

Fig. 4.17 Illustration of the model described by (4.57) and (4.58)

Given a period T, the periodicity conditions are

$$u_n(T) = u_n(0), \quad \psi_n(T) = e^{i\theta}\psi_n(0), \tag{4.60}$$

where θ is a real constant. Let $\chi_1(t; E)$ be a solution of (4.59) with period T and $\rho = 1$. In the AI limit $J = 0$, (4.57) and (4.58) have the single-site PB solution

$$u_n(t) = \begin{cases} \chi_1(t; E) & \text{if } n = 0, \\ 0 & \text{if } n \neq 0, \end{cases} \quad \psi_n(t) = \begin{cases} \exp\left[-(i/\gamma)\int_0^t u_0(s)ds\right] & \text{if } n = 0, \\ 0 & \text{if } n \neq 0. \end{cases}$$

$$\tag{4.61}$$

This solution satisfies (4.60) with $\theta = -(1/\gamma)\int_0^T u_0(s)\,ds$. There is a rigorous proof that this single-site PB and more general multi-site PBs can be continued from the AI limit to nonzero small $J > 0$, provided appropriate non-resonance and anharmonicity conditions [8]. Moreover, single-site PB solutions in (4.57) and (4.58) are numerically calculated, based on continuation with respect to J from the AI limit, and they are shown to be spectrally stable [36]. An example of numerical PB solution is shown in Fig. 4.18. The continuation with respect to J is possible up to a critical value J_c, and the solution cannot be continued beyond J_c. It is pointed out that this disappearance of the solution is accompanied with a resonance between an electronic frequency and a multiple of PB's frequency. There are some other works which numerically study PB solutions in a model slightly different from (4.57) and (4.58) [37, 38].

Finally, we point out a similarity between the concept of PB and that of DPP. It is well known that when an impurity with light mass exists in a linear lattice, a localized phonon mode can appear around the impurity. The DPP is defined as a quasi-particle composed of a DP and an impurity-induced localized phonon [13]. In both of PB and DPP, a quantum particle (electron or DP) is coupled with a localized oscillation of the lattice. The two concepts of PB and DPP are similar to each other on this point, although their mechanisms for localization are different, nonlinearity or impurity. This similarity suggests that if the lattice is a nonlinear one, there is an ILM and it could couple with a DP to form a nonlinear DPP. It may be worth exploring the possibility of nonlinear DPP.

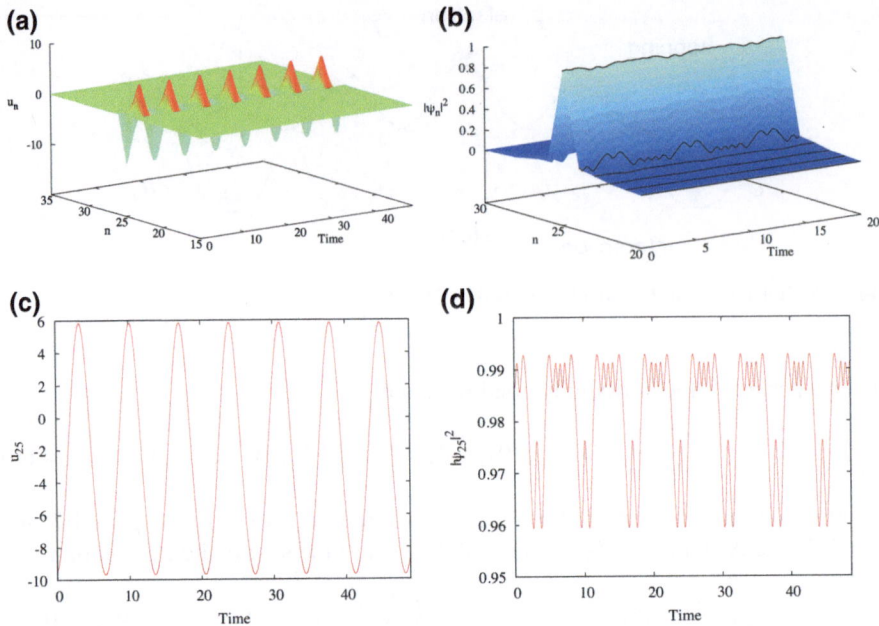

Fig. 4.18 Time evolution of polarobreather in (4.57) and (4.58): **a** displacements of oscillators $u_n(t)$; **b** probabilities $P_n \equiv |\psi_n(t)|^2$ of finding the electron at site n; **c** displacement of the center site $u_{25}(t)$; **d** probability at the center site P_{25}. The lattice size is $N = 51$, and open-end boundary conditions are used. Parameters are $J = 0.14$, $\gamma = 1$, and $T = 6.919$

4.5 Moving ILM

4.5.1 What Is Moving ILM?

In the previous sections, we discussed the standing wave type ILMs, which are also called the static ILMs. The center position of static ILM stays at a lattice site or a midpoint between two lattice sites. In addition, in various types of lattices such as FPU-β [39] and nonlinear Klein–Gordon lattice models [40], we can also observe *moving* ILMs, i.e., travelling type ILMs. Figure 4.19 shows a numerical example of a moving ILM. The center position of moving ILM travels with a certain velocity with keeping its localized structure for a long time. The shape of the moving ILM changes as its center moves: it takes the even-mode like shape when the center of ILM crosses a midpoint of lattice sites, while it takes the odd-mode like shape when the center of ILM crosses a lattice site. The moving ILM can be easily obtained by perturbing an unstable static ILM. For example, the odd mode in FPU-β lattice comes to move by adding a small perturbation.

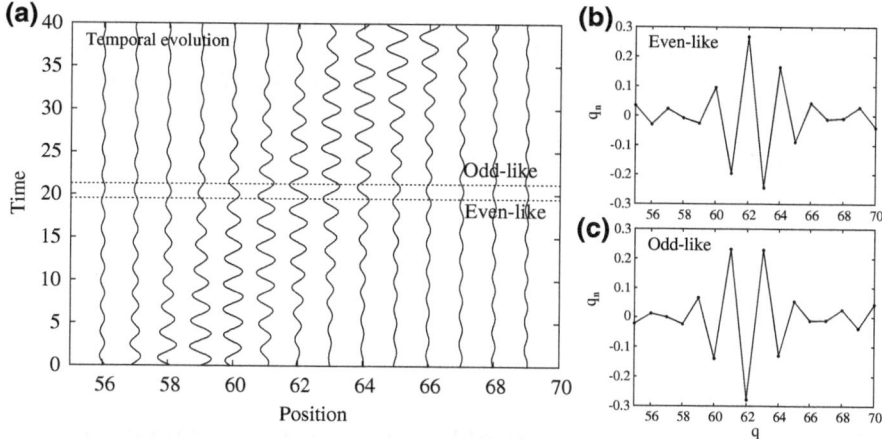

Fig. 4.19 **a** Temporal evolution of moving ILM in FPU-β lattice. *Horizontal dashed lines* indicate the instances that moving ILM takes even-like and odd like shapes. **b** and **c** are snapshots of moving ILM at these instances

4.5.2 Numerical Method for Moving ILM

Let us consider the time evolution map $\mathcal{T}: \mathbb{R}^{2n} \to \mathbb{R}^{2n}$, that is,

$$\mathcal{T}(q_1(0), q_2(0), \ldots, q_n(0), p_1(0), p_2(0), \ldots, p_n(0))$$
$$= (q_1(sT), q_2(sT), \ldots, q_n(sT), p_1(sT), p_2(sT), \ldots, p_n(sT)). \qquad (4.62)$$

In the case of the moving ILM, the following relation holds:

$$\mathcal{T}(q_1(0), q_2(0), \ldots, q_n(0), p_1(0), p_2(0), \ldots, p_n(0))$$
$$= (-1)^r(q_{1-r}(0), q_{2-r}(0), \ldots, q_{n-r}(0), p_{1-r}(0), p_{2-r}(0), \ldots, p_{n-r}(0)).$$
$$\qquad (4.63)$$

This relation means that the ILM moves r sites during s periods of its internal vibrations. We consider a case that a velocity of ILM takes a rational velocity defined by

$$v = r/s[\text{sites/period}]. \qquad (4.64)$$

A method of obtaining a numerically precise solution of moving ILM has been given in [41]. Obtaining a numerical moving ILM solution is just obtaining a initial condition that satisfies (4.63). Given internal frequency Ω and velocity v, such a solution can be calculated by Newton method as follows:

1. Prepare a precise numerical solution of the static ILM with Ω.

2. Construct an initial guess of the moving ILM by appropriately perturbing the numerical solution of the stationary ILM obtained in step 1.
3. Find a precise numerical solution of the moving ILM by Newton method with the initial guess.

Figure 4.20 shows results of numerical ILM solutions in FPU-β lattice defined by (4.4) with the periodic boundary condition. These numerical solutions are obtained by using the above-mentioned method.

It is found that the moving ILM have a finite tail in FPU-β system. This finite tail for moving ILM has been also reported in other lattice systems such as Klein–Gordon [42] and Salerno lattice [43].

4.5.3 Mobility of ILM and Symmetry of Interaction Potential

Yoshimura and Doi have suggested that the amplitude of tail of ILM highly depends on a symmetry of the interaction potential between lattice sites [41]. We define the

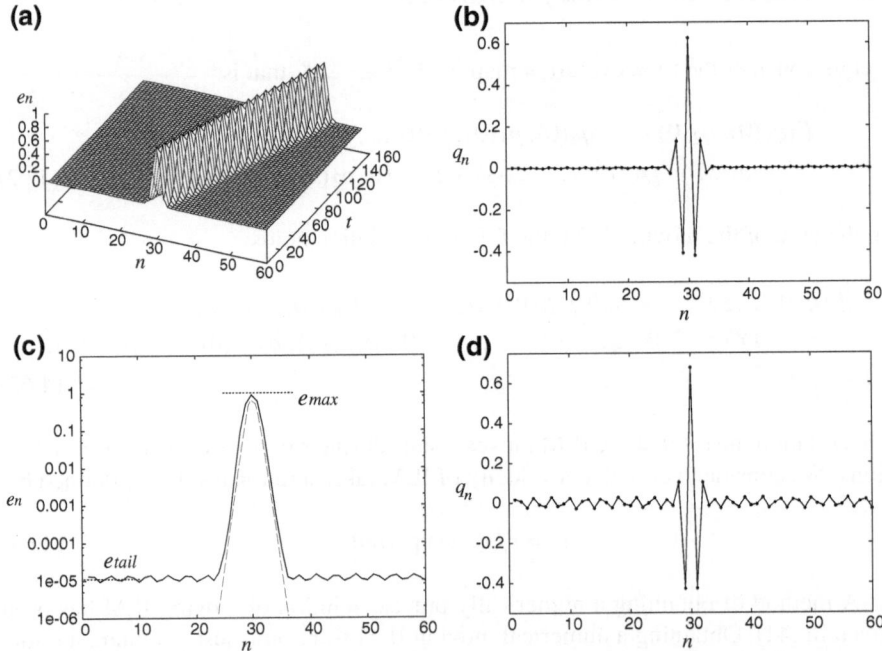

Fig. 4.20 Numerical solution of moving ILM in FPU lattice: **a** spatiotemporal plot of site energy, **b** Initial displacement in the case $\omega = 2.5$, **c** semi-logarithmic plot of site energy of static ILM (*dashed line*) and moving ILM (*solid line*) in the case $\omega = 2.5$, **d** initial displacement in the case $\omega = 2.53$. The site energy e_n is define as (4.8). Figure from [41]

symmetry of interaction potential. Consider one dimensional anharmonic lattices of the Hamiltonian

$$H = \frac{1}{2} \sum_{n=1}^{N} p_n^2 + \Phi(q_1, q_2, \ldots, q_N), \qquad (4.65)$$

where Φ is a smooth potential. We assume that Φ is invariant with respect to the uniform shift $q_n \mapsto q_n + c$, where c is a constant. We introduce the complex normal mode coordinates U_m ($m = -N/2 + 1, -N/2, \ldots, N/2$), which are defined by the transformation

$$q_n = \frac{(-1)^n}{\sqrt{N}} \sum_{m=-N/2+1}^{N/2} U_m \exp\left(i \frac{2\pi m}{N} n\right). \qquad (4.66)$$

Since Φ is invariant with respect to the uniform shift, the total momentum $\sum_{n=1}^{N} p_n = \sqrt{N} \dot{U}_{N/2}$ is a first integral of the system. Therefore, we can assume $U_{N/2} = \dot{U}_{N/2} = 0$. Let $\mathbf{U} = (U_{-N/2+1}, U_{-N/2+2}, \ldots, U_{N/2-1})$. The equations of motion in terms of U_m is given by

$$\frac{d^2 U_m}{dt^2} = -\frac{\partial \Phi(\mathbf{U})}{\partial U_{-m}}, \qquad (4.67)$$

where $m = -N/2 + 1, -N/2 + 2, \ldots, N/2 - 1$.

Consider the map $\mathcal{T}_\lambda: \mathbb{C}^{N-1} \to \mathbb{C}^{N-1}$

$$\mathcal{T}_\lambda: U_m \mapsto U_m \exp(-im\lambda), \qquad (4.68)$$

where λ is a real parameter. This map corresponds to the rotation in the U_m-space. On the other hand, in the q_n-space, this map is regarded as the translational operation.

The potential $\Phi(\mathbf{U})$ can be divided into two parts $\Phi_s(\mathbf{U})$ and $\Phi_a(\mathbf{U})$. The former part Φ_s is a symmetric part with respect to map $\mathcal{T}_\lambda$ for any λ, that is,

$$\Phi_s(\mathcal{T}_\lambda \mathbf{U}) = \Phi_s(\mathbf{U}). \qquad (4.69)$$

The latter part Φ_a is an asymmetric part defined as

$$\Phi_a = \Phi - \Phi_s. \qquad (4.70)$$

It has been suggested that the finite tail of the moving ILM appears due to the existence of this asymmetric part of potential [41]. When we consider the symmetric lattice that contains only the symmetric part of potential, it is expected that the moving ILM without the tail can be obtained.

In general lattices that contain the asymmetric part of potential, a moving ILM that is produced by arbitrary perturbation gradually loses its energy. In the symmetric lattice, in contrast, it has been numerically found that such a moving ILM does not lose its energy and travels with a constant velocity [44]. In addition, numerically exact moving ILM solutions have been obtained in a four particle symmetric lattice [45].

4.5.4 Collision Dynamics of ILM

Mobility of ILM leads to the possibility of interaction between ILMs in the system. Let us consider the modulational instability of zone boundary mode. As a result of modulational instability of zone boundary mode, many moving ILMs are excited in the system [46]. Then excited ILMs move and interact randomly. The number of ILMs decreases as time passes. Finally only one ILM survives in the system. Randomly moving ILMs in the process are called chaotic breathers. This process indicates that two moving ILMs exchange their energies during interaction. This is quite different from collisions of two solitons that preserve their energies after interaction.

Collision between two moving ILMs has been systematically investigated in FPU-β system [47]. ILMs collide inelastically depending on their energies, phase deference of internal vibration, and distance. Heuristic model equations which describe collision of ILMs are given as follows:

$$\frac{dE_1}{dt} = -\frac{A}{2} \sin \Delta\Phi \exp\left(-\frac{\bar{x}}{L}\right), \qquad (4.71)$$

$$\frac{dE_2}{dt} = \frac{A}{2} \sin \Delta\Phi \exp\left(-\frac{\bar{x}}{L}\right), \qquad (4.72)$$

$$\frac{d\Delta\Phi}{dt} = F(E_2) - F(E_1), \qquad (4.73)$$

$$\frac{d^2\bar{x}}{dt^2} = -B \cos \Delta\Phi \exp\left(-\frac{\bar{x}}{L}\right), \qquad (4.74)$$

where A, B, and L are constants determined from the numerical results, E_1 and E_2 are energies of two ILMs, $\bar{x}$ is distance between two ILMs, $\Delta\Phi$ is phase difference of the two ILMs. $F(E)$ is the function that describes a relation between energy and internal frequency of ILM. Figure 4.21a–d shows dynamics of two ILMs with different energy. Four figures indicate the results with different initial phase differences. Dynamics drastically depends on the initial condition of phase difference between two ILMs even when two ILM have the same energy. Figure 4.21e shows range of resultant energy of after interaction. Exchanged energy varies depending on the state of ILM before interaction. In many cases, difference of energies of two ILMs after collision becomes larger than that before collision. Therefore the energy of one ILM

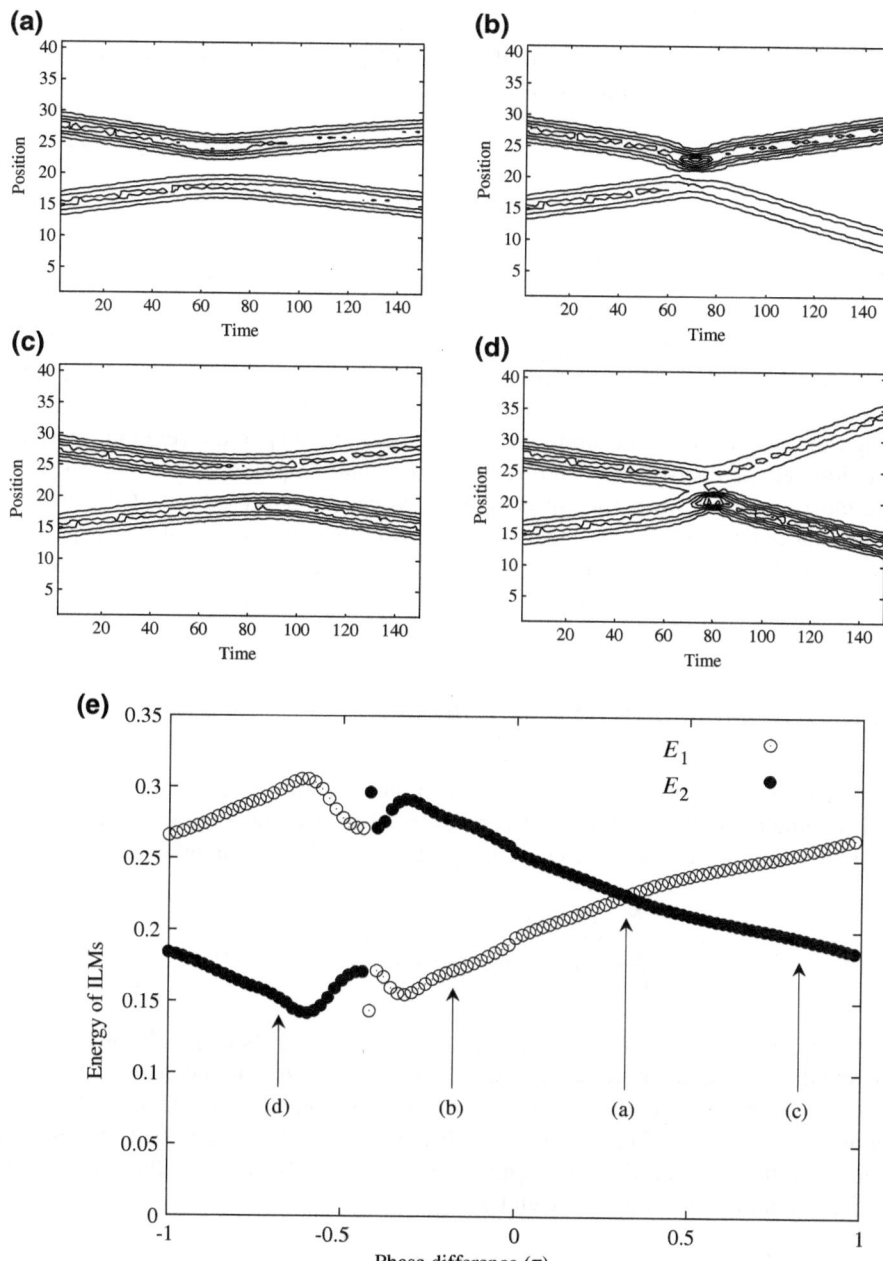

Fig. 4.21 Collision of two ILMs: energy of ILMs are $E_1 = 0.2$ and $E_2 = 0.25$. Phase differences at instance of collision are **a** $\pi/2 < \Delta\Phi < \pi$, **b** $0 < \Delta\Phi < \pi/2$, **c** $-\pi < \Delta\Phi < \pi/2$, and **d** $-\pi/2 < \Delta\Phi < 0$. Resultant energy of two ILMs after interaction is shown in (**e**). Labels (a)–(d) in graph corresponds to (**a**)–(**d**)

after collision becomes larger than the energies of two ILMs before collision, in many cases. This is consistent with the result of [46] that one of the excited ILMs gains almost all energy from the other ILMs.

4.6 Numerical Simulations of ILM in Real Systems

It is well known that ILMs ubiquitously exist in large class of nonlinear lattices. Therefore it is expected that ILMs can be excited in real systems. One typical example is ILMs as atomic vibrations in crystals. In the microscopic view of crystals, atoms arrange periodically in space. Interaction potentials between atoms are generally described as nonlinear functions of their distance. Nonlinearity becomes dominant when the displacement of atoms becomes large. The dispersion relation, that is a relation between wavelength and frequency in small amplitude vibration, is one of the most important properties of crystals. ILM can be observed out of the phonon band that is bounded by the maximum and minimum frequency of the dispersion curve. In this section, we show some numerical results of excitation of ILMs in crystals by molecular dynamic (MD) simulations.

4.6.1 Excitation of ILM in Graphene

Graphene is a two dimensional crystal that has honeycomb structure of carbon atoms. The interaction between atoms is described by a heuristic model potential for hydrocarbon system proposed by Brenner [48]. The model Hamiltonian is

$$H = \sum_{i=1}^{N} \frac{\mathbf{p}_i^2}{2m} + \frac{1}{2} \sum_{i}^{N} \sum_{j \neq i}^{N} \sum_{k \neq i,j}^{N} \Phi_{ijk}(r_{ij}, r_{ik}, \theta_{ijk}), \tag{4.75}$$

where i, j, and k are the index of atoms, m is the mass of carbon, $\mathbf{p}_i \in \mathbb{R}^3$ is the linear momentum of i-th atom, $r_{ij} = |\mathbf{q}_j - \mathbf{q}_i|$ and $r_{ik} = |\mathbf{q}_k - \mathbf{q}_i|$ are the distance between two atoms, $\mathbf{q}_i \in \mathbb{R}^3$ is the position vector of i-th atoms, θ_{ijk} is the angle between bonds $i - j$ and $i - k$. Brenner potential Φ_{ijk} reproduces the honeycomb structure of carbon atoms. The equilibrium length between atoms is 0.145 nm in the simulations.

The equation of motion is given by

$$\dot{\mathbf{q}}_i = \frac{\partial H}{\partial \mathbf{p}_i} \tag{4.76}$$

$$\dot{\mathbf{p}}_i = -\frac{\partial H}{\partial \mathbf{q}_i}. \tag{4.77}$$

In the MD simulation, a thermal bath is usually connected to the system in order to maintain the thermally equilibrium state. However, in our investigation of ILM, no

Fig. 4.22 Snapshot of vibration of ILM in graphene. Center of two neighboring atoms vibrate with large amplitude along a bond

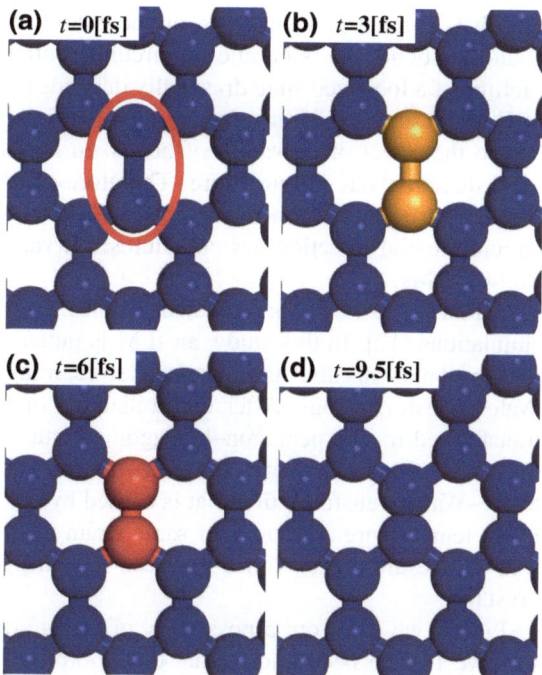

thermal bath is connected, since we consider that energy of the system is conserved through the simulation.

We search a numerical solution of ILM in graphene system by Newton method. Figure 4.22 shows snapshots of the numerical solution of the ILM in graphene [49]. Two atoms vibrate with large amplitudes. The surrounding atoms do not vibrate noticeably. Therefore ILM is successfully obtained. Direction of vibration is along the bond of atoms. This structure of ILM in graphene is also observed in MD simulation. In this case, ILMs are spontaneously excited from modulational instability of zone boundary mode of graphene [50]. ILMs are excited in three directions that correspond to three directions of the bonds between neighboring atoms.

4.6.2 Excitation of ILM in CNT

Carbon nanotube (CNT) is a three-dimensional structure of carbon atoms that is made by rolling up graphene. MD simulations of excitation of ILMs in CNTs have been also performed [51]. In MD simulation of modulational instability from initial displacement, ILMs are excited as same as the case of graphene. However, direction of oscillation of ILM is limited to specific direction of bond. In the zigzag CNTs, direction of oscillation of ILM is along a bond that is parallel to axial direction of the CNT. In the armchair CNTs, on the other hand, direction of oscillation of ILM is along a bond that is parallel to circumferential direction of the CNT. Moreover lifetime of a localized state drastically depends on the structures of CNTs. Lifetime of ILMs in zigzag CNTs is longer than that in armchair CNTs. This difference is due to the effect of curvature of bonds. In the zigzag CNTs, the bond along the axial direction has no curvature. Therefore the ILM in this direction is same as the ILM in graphene. In the armchair CNTs, on the other hand, bond along the circumferential direction has curvatures. Curvature of bond affects stability of the excited ILMs.

Shimada et al. have investigated the mechanical effect of ILM in CNTs by MD simulations [12]. In this study, an ILM is putted on the strained CNT at first. The ILM becomes unstable and then leads to rearrangement of atoms known as Stone–Wales transformations. After unstabilization of ILM, four hexagon structures are transformed to two pentagon–heptagon structures. Usually, Stone–Wales transformation is occurs under very high temperature and high-strained conditions. However Stone–Wales transformation that is caused by ILM occurs the condition with much lower temperature and lower or same strain than conditions in usual cases. Therefore this result suggests that ILM can be a trigger of rearrangement of atoms in crystals.

Finally we point out a possibility of change of atomic structure in crystals due to ILM. It have been known that the amorphous and crystalline states of chalcogenide glass have different optical properties. Therefore chalcogenide glass has been investigated as phase change materials. The phase change of chalcogenide glass occurs by an irradiation of light. Recently it has been reported that fast transition between the amorphous and crystalline states in small region is realized by irradiation of short time laser pulse [52, 53]. This transition process is not heating process since an irradiation time is too short to heat the material. It is expected that ILM is excited by laser pulse and that ILM plays a trigger of structure change in atomic scales. More exploring the possibility of nonlinear dynamics in atomic scale is expected.

4.7 Energy Localization in Experiments

By reflecting the ubiquitousness of energy localizations, there have been reported many experimental results. To observe ILMs in experiment, nonlinearity and homogeneity in lattice are necessary at least. For this reason, the photolithography technique is frequently used to create nonlinear homogeneous lattices. Therefore, experimental studies of ILM have often been done in nano-/micro-scale systems. In this section, we pick several examples up from the nano-/micro-scale systems. In addition, a macro-mechanical system is also mentioned.

4.7.1 Coupled Flexible Beams

Flexible beam structure is widely used in engineering, especially in micro-engineering in which the beam is used as a linear/nonlinear resonator. The photolithography technique enables us to fabricate micro-/nano-meter scale oscillators on a semi-conductor wafer. An advantage of the technique is high accuracy of fabrication which means the easiness of producing homogeneous structures in which oscillators have almost same characteristics each other. Using this advantage, Sato et al. created the micro-cantilever array consisting of several hundred cantilevers, the overhang, and the support (see Fig. 4.23) [6]. The cantilevers act as nonlinear oscillators at which the nonlinearity mainly comes from the geometric nonlinearity of flexible beam. In the array, short and long cantilevers are alternately arranged so that the zone boundary mode can be excited by an external uniform excitation. The overhang is a thin film which is bended by cantilever's deformation. That is, the overhang gives rise to coupling effect between adjacent cantilevers. The structure mentioned above is attached on the support which is a thick plate.

Intrinsic localized modes were successfully observed when the external vibration was applied to the cantilevers and the overhang via the support [6]. They used a piezo film to make the vibration. The frequency is chirped up from that of the zone boundary mode for generating ILMs through the modulational instability. As shown in Fig. 4.24, several ILMs are simultaneously excited and they keep their energy concentration until the external vibration is turned off. It is confirmed experimentally and numerically that three or four cantilevers having large amplitudes compose each ILM [55]. That is, the energy of the observed ILMs is concentrated only in a few sites. Interestingly, their experiments also show ILMs wandering reciprocally. Although they cannot survive for a long period, the energy concentration is kept while they

Fig. 4.23 Micro-cantilever array fabricated by Sato et al. Figures from [54]

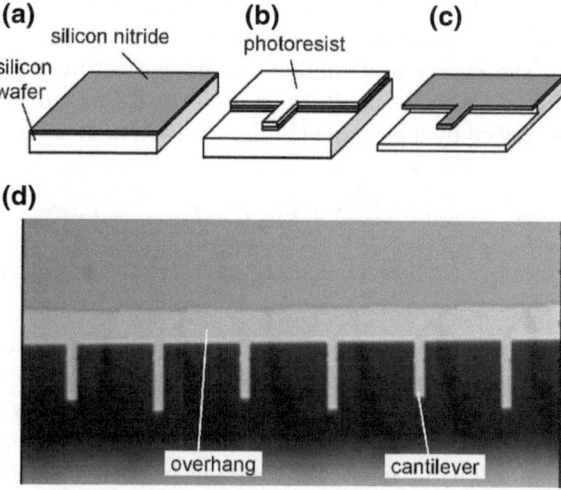

Fig. 4.24 Observation of ILMs in the micro-cantilever array. *Dark regions* indicate where cantilevers oscillate with large amplitude. The external vibration is turned on at $t = 0$ and is turned off at $t \simeq 50$ ms. Figures from [6]

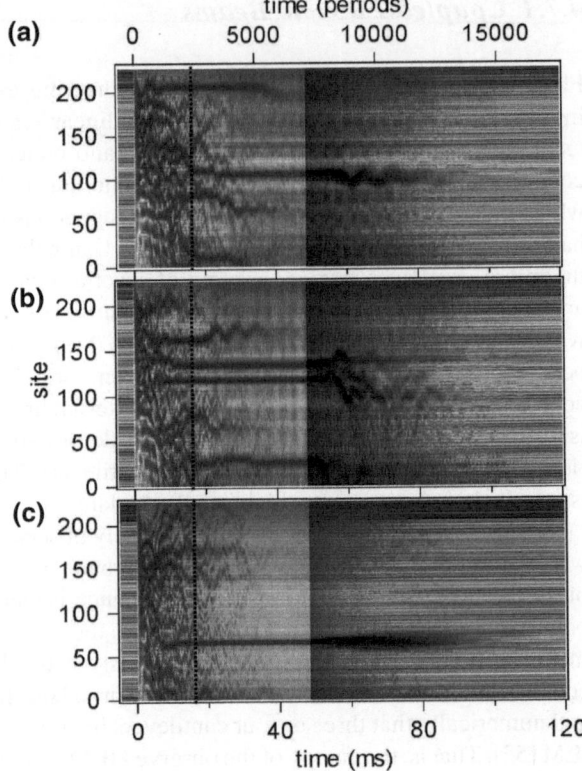

are moving. The result implies the existence of moving ILM in real system. In fact, Sato et al. succeeded to manipulate ILM by using a locally induced impurity [56]. A static ILM is attracted or repelled with the impurity.

The equation of motion can be derived by focusing on the first mode vibration of Euler–Bernoulli beam. Motion of coupled cantilevers is then approximately described by the coupled ordinary differential equation

$$\frac{d^2 x_i}{dt^2} + k_{2Oi}x_i + k_{4Oi}x_i^3 + k_{2I}(2x_i - x_{i-1} - x_{i+1})$$
$$+ k_{4I}\left\{(x_i - x_{i-1})^3 + (x_i - x_{i+1})^3\right\} = 0, \tag{4.78}$$

where x_n represents the displacement of the tip of the nth cantilever from an equilibrium position. The linear on-site and coupling coefficients are represented by k_{2O} and k_{2I}, respectively. In this equation, cubic nonlinearity of restoring force is considered for both on-site and coupling. The coefficients k_{4O} and k_{4I} denote magnitudes of the nonlinearity in on-site and coupling, respectively. Sato et al. also showed that numerical simulations using (4.78) or a similar equation well reproduce the corresponding experimental result. Therefore, the coupled cantilever array can be classified into the mixed model of the FPU lattice and the NKG lattice. As mentioned in Sect. 4.3.3, the stability of ILM can be exchanged owing to the ratio of nonlinear coefficients such as k_{4O} and k_{4I} in (4.78) [23]. By using the fact, a manipulation method called "Capture and Release" manipulation is proposed numerically [57]. In this method, the ratio of nonlinear coefficients are uniformly varied to switch the stability of ILM. If the stability of ILM is switched, a stable ILM loses its stability and begins to move. The moving ILM can be captured if the stability is switched again at an appropriate timing.

Another example of coupled flexible beam is a macro-mechanical system having magnetic interactions [7, 58]. Figure 4.25 shows the experimental equipment of the system. Several-centimeter-cantilevers are arranged one direction as well as the micro-cantilever array. These cantilevers are coupled by the coupling rod which is an elastic rod. The main difference from the micro-cantilever array is how nonlinearity is induced to restoring force against displacement of cantilevers. In the macro-mechanical system, magnetic force is used to make the nonlinearity because the geometric nonlinearity is rather weak comparing with that of magnetic force. In addition, the magnetic force has an advantage of adjustable such that the magnetic flux depends on the current flowing in electromagnet. In this system, electromagnets are placed beneath permanent magnets which are attached to the tip of cantilevers. Therefore, the on-site nonlinearity is individually adjustable. On the other hand, the coupling rod does not deform largely. Then, the nonlinearity in the coupling is negligible. The macro-mechanical system is classified into NKG lattice.

Figure 4.26a shows generation and manipulation of ILM in the macro-mechanical system. As shown in the figure, an ILM is generated at $n = 4$ and survives until $t = 5$ s. Generating ILM is realized by adding an impurity, which is created by decreasing the current flowing the electromagnet of fourth site I_{EM} for a while. The same impurity also attracts an ILM standing at neighbor site as shown in Fig. 4.26b. The ILM is initially excited at $n = 5$. Then the ILM is attracted with the impurity.

Fig. 4.25 Macro-mechanical cantilever array having tunable on-site potential. The voice coil motor vibrate all the cantilevers simultaneously through the support which is a rigid bar of Aluminum

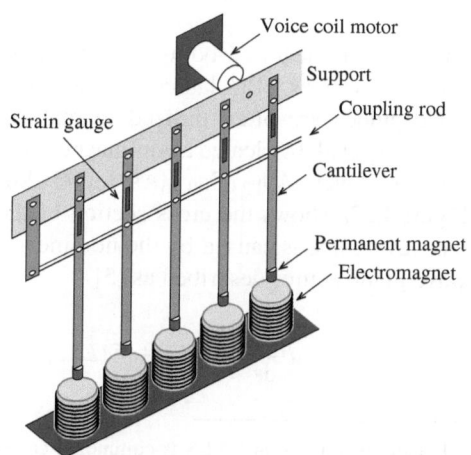

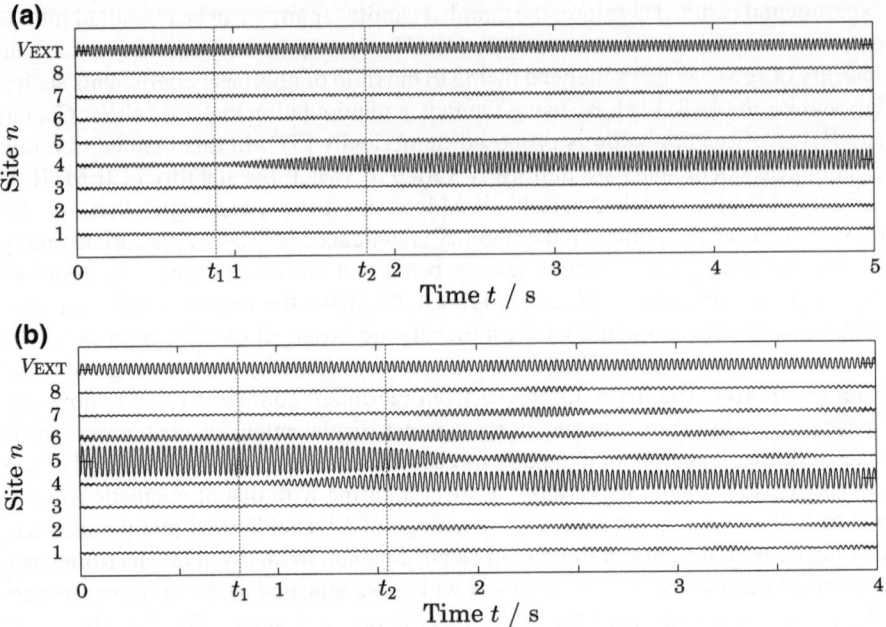

Fig. 4.26 a An ILM is generated at $n = 4$ by adding an impurity. **b** An ILM which initially stands at $n = 5$ is attractively manipulated to $n = 4$ with the impurity. For both cases, the impurity at $n = 4$ is added at $t = t_1$ and is removed at $t = t_2$

Finally the ILM is trapped at $n = 4$. Namely, the ILM is manipulated from $n = 4$ to $n = 5$.

4.7.2 Nonlinear Optic Wave Guides

Discrete nonlinear Schrödinger (DNLS) lattice is also investigated by many researchers as well as FPU lattice or NKG lattice.[1] For experiments, nonlinear optic wave guides are often utilized to realize nonlinear lattice described by DNLS. Eisenberg et al. fabricated a nonlinear optic wave guide array made of $Al_{0.18}Ga_{0.82}As$ for core layer, $Al_{0.24}Ga_{0.76}As$ for cladding layer, and GaAs for substrate [5]. Figure 4.27a shows the cross section of the wave guide array. The nonlinearity in the wave guide is caused by the nonlinear Kerr effect. The electric field E_n in each wave guide is thus described as [5]

$$i\frac{dE_n}{dz} + \beta E_n + C(E_{n-1} + E_{n+1}) + \gamma |E_n|^2 E_n = 0, \qquad (4.79)$$

---

[1] Localized solution in DNLS is commonly called *discrete soliton* instead of intrinsic localized mode or discrete breather

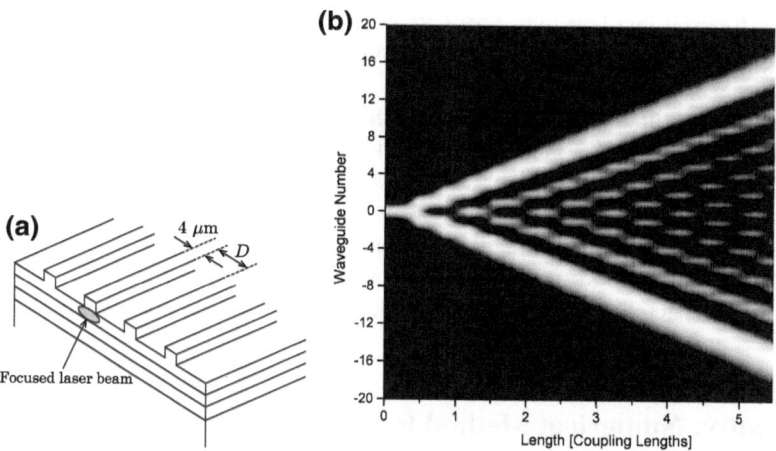

Fig. 4.27 **a** Optic wave guide array fabricated by Eisenberg. This *figure* is drawn based on the Fig. 2 in [5]. **b** Propagation of light when the nonlinearity is negligibly small. Figure from [5]

where C is a coupling coefficient which is determined by the distance between adjacent wave guides D, and γ denotes the magnitude of cubic nonlinearity caused by the nonlinear Kerr effect.

When the electric field is small enough, the light propagates linearly as shown in Fig. 4.27b. However, the focused light does not spread when the power of the incident light beam is strong enough. Figure 4.28 shows how the distribution of light on the facet changes with the peak power of the incident light. The light is clearly concentrated in Fig. 4.28c. The dynamics of ILM in the wave guide array is also investigated by varying the initial position of ILM [59]. In this system, site-centered modes are stable whereas bond-centered modes are unstable.

Experimental studies of ILM have been limited in one dimensional lattice because of the difficulty of fabrication and measurement. However, in 2003, the limitation has been broken by Fleischer et al. They realized a two dimensional optic wave guide array using the optic induction technique in which interference pattern of two plane waves are used [60]. As well as the case of one dimensional wave guide array, the

Fig. 4.28 Observation of ILM on the facet of the wave guide array. The peak power is **a** 70 W, **b** 320 W, **c** 500 W. Figure from [5]

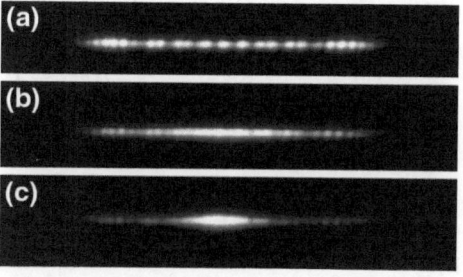

initially focused incident light keeps its concentration if the nonlinearity is strong enough. The feasibility of experiment in three dimensional lattice is also mentioned in [61].

Even in one dimensional lattice, ILM shows intriguing behaviors such as ratcheting behavior [62] and spontaneously walking [63]. For the ratcheting behavior, an ILM moves unidirectionally like a ratchet mechanism. This implies the possibility of control of ILM in optical wave guide array. For the spontaneously walking, an ILM wanders at random owing to the lost of stability by changing the coupling constant. This shows the possibility of creating moving ILM from static ILM. These facts allow us to expect the realization of applications using ILM in *micro/nanophotonics*.

Appendix: Numerical Method for Computing ILM

ILM solutions for different periods form a one-parameter family of closed orbits in phase space, as mentioned in Sect. 4.2.3. Given a period T, an ILM solution with the period T corresponds to one of these closed orbits. The Newton method is a useful tool to find such a closed orbit. We describe this method according to [40].

Let us consider (4.2) and its solution $z(t) \equiv (q(t), p(t))^t \in \mathbb{R}^{2N}$, where superscript t stands for the transposition, $q = (q_1, \ldots, q_N)$, and $p = (p_1, \ldots, p_N)$. Integration of (4.2) over a time interval $[0, t]$ with the initial condition $z(0)$ defines the time evolution map $\mathcal{F}_t : \mathbb{R}^{2N} \to \mathbb{R}^{2N}$, $z(0) \mapsto z(t)$. In this notation, the initial point $z(0)$ of a T-periodic ILM is a fixed point of the map $\mathcal{F}_T$, i.e., $\mathcal{F}_T[z(0)] = z(0)$. An iteration procedure can be used for finding a fixed point of $\mathcal{F}_T$.

If z_0 is a point close to the desired fixed point of $\mathcal{F}_T$, then a small variation Δ results in $\mathcal{F}_T[z_0 + \Delta] \simeq \mathcal{F}_T[z_0] + D\mathcal{F}_T \cdot \Delta$, where $D\mathcal{F}_T$ is the Jacobian matrix of $\mathcal{F}_T$ evaluated at the point z_0. Let $z(t; z_0)$ denote the solution of (4.2) such that $z(0; z_0) = z_0$. The variational equation of (4.2) along $z(t; z_0)$ can be written as follows:

$$\dot{\Phi}(t) = J H_{zz}(z(t; z_0)) \cdot \Phi(t), \tag{4.80}$$

where J is the symplectic matrix given by (4.48) and $\Phi(t)$ is the fundamental matrix satisfying the initial condition $\Phi(0) = I$, where I represents the $2N \times 2N$ identity matrix. The matrix $D\mathcal{F}_T$ is just given by $D\mathcal{F}_T = \Phi(T)$, and it has the form

$$D\mathcal{F}_T = \begin{pmatrix} A & B \\ C & D \end{pmatrix}, \tag{4.81}$$

where A, B, C, D are $N \times N$ matrices.

The problem of finding the desired fixed point is equivalent to minimizing the distance $\|\mathcal{F}_T[z_0 + \Delta] - (z_0 + \Delta)\|^2$ with respect to Δ. In order to approximately perform this minimization, we minimize $\|\mathcal{F}_T[z_0] + D\mathcal{F}_T \cdot \Delta - (z_0 + \Delta)\|^2$. The vector Δ consists of N position and N momentum components, i.e., $\Delta = (\delta_q, \delta_p)^t \in \mathbb{R}^{2N}$.

We perform the minimization only with respect to the position components δ_q, since the ILM solution can be assumed to have zero momentum at $t = 0$. That is, we take $\Delta = (\delta_q, 0)^t$. Minimization with respect to δ_q leads to

$$\delta_q = -(A^*A + C^*C)^{-1}[A^*, C^*] \cdot \{\mathcal{F}_T[z_0] - z_0\}, \qquad (4.82)$$

giving an improved approximation $z_0' = z_0 + (\delta_q, 0)^t$. In (4.82), $\mathcal{F}_T[z_0] - z_0$ is a vector of dimension $2N$, $[A^*, C^*]$ is an $N \times 2N$ matrix formed by placing matrices A^* and C^* adjacently in a new nonsquare matrix, $(A^*A + C^*C)^{-1}$ is an $N \times N$ matrix, and the asterisk denotes conjugation. If we repeatedly apply this procedure, we obtain an accurate solution for the ILM with the given period T.

A good initial approximation for the ILM solution is necessary to successfully apply the above Newton method. A systematic method for obtaining an initial approximation has been proposed in [16], based on the AI limit concept. For the FPU lattice (4.4), a different approach is also possible. One can use (4.32) as a good approximation to compute a T-periodic ILM with $T \simeq \pi$, i.e., in the case of $\Omega \simeq \omega_{max} = 2$. It is also possible to compute an ILM with T small ($\Omega \gg \omega_{max}$) by starting with an appropriately chosen period $T_0 \simeq \pi$ and decreasing the period by a small step δT. That is, first, we compute the ILM with the initial period T_0, using (4.32) as an approximation. Then, we use the numerical ILM solution at period T_0 as the initial approximation to compute the ILM solution at period $T_0 - \delta T$. A similar procedure is repeated until the period reaches a desired one.

It sometimes happens that the lattice model has additional first integrals besides the energy integral. For example, the total momentum $\sum_{n=1}^{N} p_n$ is conserved in FPU lattices with the periodic boundary conditions. In such a case, the above minimization problem with respect to Δ becomes degenerated, and thus one has to carry out the calculation under appropriate constraints on Δ in order to remove this degeneracy.

References

1. S. Takeno, K. Kisoda, A.J. Sievers, Intrinsic localized vibrational modes in anharmonic crystals: stationary modes. Prog. Theor. Phys. Suppl. **94**, 242–269 (1988)
2. A.J. Sievers, S. Takeno, Intrinsic localized modes in anharmonic crystals. Phys. Rev. Lett. **61**, 970–973 (1988)
3. E. Trias, J.J. Mazo, T.P. Orlando, Discrete breathers in nonlinear lattices: experimental detection in a Josephson array. Phys. Rev. Lett. **84**, 741–744 (2000)
4. P. Binder, D. Abraimov, A.V. Ustinov, S. Flach, Y. Zolotaryuk, Observation of breathers in Josephson ladders. Phys. Rev. Lett. **84**, 745–748 (2000)
5. H.S. Eisenberg, Y. Silberberg, R. Morandotti, A.R. Boyd, J.S. Aitchison, Discrete spatial optical solitons in waveguide arrays. Phys. Rev. Lett. **81**, 3383–3386 (1998)
6. M. Sato, B.E. Hubbard, A.J. Sievers, B. Ilic, D.A. Czaplewski, H.G. Craighead, Observation of locked intrinsic localized vibrational modes in a micromechanical oscillator array. Phys. Rev. Lett. **90**, 044102 (2003)
7. M. Kimura, T. Hikihara, Coupled cantilever array with tunable on-site nonlinearity and observation of localized oscillations. Phys. Lett. A **373**, 1257–1260 (2009)

8. S. Aubry, Breathers in nonlinear lattices: existence, linear stability and quantization. Physica D **103**, 201–250 (1997)
9. S. Flach, C. Willis, Discrete breathers. Phys. Rep. **295**, 181–264 (1998)
10. S. Aubry, Discrete breathers: localization and transfer of energy in discrete Hamiltonian nonlinear systems. Physica D **216**, 1–30 (2006)
11. S. Flach, A.V. Gorbach, Discrete breathers—advances in theory and applications. Phys. Rep. **467**, 1–116 (2008)
12. T. Shimada, D. Shirasaki, T. Kitamura, Stone–Wales transformation triggered by intrinsic localized modes in carbon nanotubes. Phys. Rev. B **81**, 035401 (2010)
13. M. Ohtsu, *Dressed Photon* (Asakura, Tokyo, 2013). (in Japanese)
14. E. Fermi, J. Pasta, S. Ulam, in *Collected Papers of E. Fermi*, ed. by E. Segré (University of Chicago Press, Chicago, 1965)
15. J.B. Page, Asymptotic solutions for localized vibrational modes in strongly anharmonic periodic systems. Phys. Rev. B **41**, 7835–7838 (1990)
16. J.L. Marín, S. Aubry, Breathers in nonlinear lattices: numerical calculation from the anticontinuous limit. Nonlinearity **9**, 1501–1528 (1996)
17. B. Dey, M. Eleftheriou, S. Flach, G.P. Tsironis, Shape profile of compact-like discrete breathers in nonlinear dispersive lattice systems. Phys. Rev. E **65**, 017601 (2001)
18. Y. Doi, A. Nakatani, Intrinsic localized mode as in-plane vibration in two-dimensional Fermi-Pasta-Ulam lattices. Nonlinear Theory Its Appl. IEICE **3**, 67–76 (2012)
19. Y.A. Kosevich, New soliton equation and exotic localized modes in anharmonic lattices. Phys. Lett. A **173**, 257–262 (1993)
20. Y.A. Kosevich, S. Lepri, Modulational instability and energy localization in anharmonic lattices at finite energy density. Phys. Rev. B **61**, 299–307 (2000)
21. D.D. Holm, J.E. Marsden, T. Ratiu, A. Weinstein, Nonlinear stability of fluid and plasma equilibria. Phys. Rep. **123**, 1–116 (1985)
22. K.W. Sandusky, J.B. Page, K.E. Schmidt, Stability and motion of intrinsic localized modes in nonlinear periodic lattices. Phys. Rev. B **46**, 6161–6168 (1992)
23. M. Kimura, T. Hikihara, Stability change of intrinsic localized mode in finite nonlinear coupled oscillators. Phys. Lett. A **372**, 4592–4595 (2008)
24. S. Flach, Existence of localized excitations in nonlinear Hamiltonian lattices. Phys. Rev. E **51**, 1503–1507 (1995)
25. S. Aubry, G. Kopidakis, V. Kadelburg, Variational proof for hard discrete breathers in some classes of Hamiltonian dynamical systems. Discrete Continuous Dyn. Syst. B **1**, 271–298 (2001)
26. G. James, Centre Manifold reduction for quasilinear discrete systems. J. Nonlinear Sci. **13**, 27–63 (2003)
27. K. Yoshimura, in *Proceedings of the 2013 International Symposium on Nonlinear Theory and Its Applications*, Existence and stability of discrete breathers in Fermi-Pasta-Ulam type lattices, pp. 274–277, 2013
28. S. Aubry, G. Abramovici, Chaotic trajectories in the standard map. The concept of anti-integrability. Physica D **43**, 199–219 (1990)
29. R.S. MacKay, S. Aubry, Proof of existence of breathers for time-reversible or Hamiltonian networks of weakly coupled oscillators. Nonlinearity **7**, 1623–1643 (1994)
30. J.F.R. Archilla, J. Cuevas, B. Sánchez-Rey, A. Alvarez, Demonstration of the stability or instability of multibreathers at low coupling. Physica D **180**, 235–255 (2003)
31. V. Koukouloyannis, P.G. Kevrekidis, On the stability of multibreathers in Klein–Gordon chains. Nonlinearity **22**, 2269–2285 (2009)
32. D. Pelinovsky, A. Sakovich, Multi-site breathers in Klein–Gordon lattices: stability, resonances, and bifurcations. Nonlinearity **25**, 3423–3451 (2012)
33. K. Yoshimura, Stability of discrete breathers in nonlinear Klein–Gordon type lattices with pure anharmonic couplings. J. Math. Phys. **53**, 102701 (2012)
34. R. Livi, M. Spicci, R.S. MacKay, Breathers on a diatomic FPU chain. Nonlinearity **10**, 1421–1434 (1997)

35. K. Yoshimura, Existence and stability of discrete breathers in diatomic Fermi-Pasta-Ulam type lattices. Nonlinearity **24**, 293–317 (2011)
36. G. Kalosakas, S. Aubry, Polarobreathers in a generalized Holstein model. Physica D **113**, 228–232 (1998)
37. M.A. Fuentes, P. Maniadis, G. Kalosakas, K.Ø. Rasmussen, A.R. Bishop, V.M. Kenkre, YuB Gaididei, Multipeaked polarons in soft potentials. Phys. Rev. E **70**, 025601R (2004)
38. J. Cuevas, P.G. Kevrekidis, D.J. Frantzeskakis, A.R. Bishop, Existence of bound states of a polaron with a breather in soft potentials. Phys. Rev. B **74**, 064304 (2006)
39. K. Hori, S. Takeno, Moving self-localized modes for the displacement field in a one-dimensional lattice system with quartic anharmonicity. J. Phys. Soc. Jpn. **51**, 2186–2189 (1992)
40. D. Chen, S. Aubry, G.P. Tsironis, Breather mobility in discrete ϕ^4 nonlinear lattices. Phys. Rev. Lett. **77**, 4776–4779 (1996)
41. K. Yoshimura, Y. Doi, Moving discrete breathers in nonlinear lattice: resonance and stability. Wave Motion **45**, 83–99 (2007)
42. S. Aubry, T. Cretegny, Mobility and reactivity of discrete breathers. Physica D **119**, 34–46 (1998)
43. J. Gómez-Gardeñes, F. Falo, L.M. Floría, Mobile localization in nonlinear Schrödinger lattices. Phys. Lett. A **332**, 213–219 (2004)
44. Y. Doi, K. Yoshimura, in Proceedings of the International Symposium on Nonlinear Theory and Its Applications, Constructing a lattice model supporting highly mobile discrete breathers, 2014
45. Y. Doi, K. Yoshimura, Translational asymmetry controlled lattice and numerical method for moving discrete breather in four particle system. J. Phys. Soc. Jpn. **78**, 034401
46. T. Cretegny, D. Dauxois, S. Ruffo, A. Torcini, Localization and equipartition of energy in the β-FPU chain: chaotic breathers. Physica D **121**, 109–126 (1998)
47. Y. Doi, Energy exchange in collisions of intrinsic localized modes. Phys. Rev. E **68**, 066608 (2003)
48. D.W. Brenner, Empirical potential for hydrocarbons for use in simulating the chemical vapor deposition of diamond films. Phys. Rev. B **42**, 9458 (1990)
49. Y. Doi, A. Nakatani, Numerical study on unstable perturbation of intrinsic localized modes in graphene. J. Solid Mech. Mater. Eng. **7**, 540–552 (2013)
50. Y. Yamayose, Y. Kinoshita, Y. Doi, A. Nakatani, T. Kitamura, Excitation of intrinsic localized modes in a graphene sheet. EPL **80**, 40008 (2007)
51. T. Shimada, D. Shirasaki, Y. Kinoshita, Y. Doi, A. Nakatani, T. Kitamura, Influence of nonlinear atomic interaction on excitation of intrinsic localized modes in carbon nanotubes. Physica D **239**, 407–413 (2010)
52. M. Hase, Y. Miyamoto, J. Tominaga, Ultrafast dephasing of coherent optical phonons in atomically controlled GeTe/Sb$_2$2Te$_3$ superlattices. Phys. Rev. B **79**, 174112 (2009)
53. K. Makino, J. Tominaga, M. Hase, Ultrafast optical manipulation of atomic arrangements in chalcogenide alloy memory materials. Opt. Express **19**, 1260–1270 (2011)
54. M. Sato, B.E. Hubbard, A.J. Sievers, Colloquium: nonlinear energy localization and its manipulation in micromechanical oscillator arrays. Rev. Mod. Phys. **78**, 137–157 (2006)
55. M. Sato, B.E. Hubbard, L.Q. English, A.J. Sievers, B. Ilic, D.A. Czaplewski, H.G. Craighead, Study of intrinsic localized vibrational modes in micromechanical oscillator arrays. Chaos **13**, 702–715 (2003)
56. M. Sato, B.E. Hubbard, A.J. Sievers, B. Ilic, H.G. Craighead, Optical manipulation of intrinsic localized vibrational energy in cantilever arrays. Europhys. Lett. **66**, 318–323 (2004)
57. M. Kimura, T. Hikihara, Capture and release of traveling intrinsic localized mode in coupled cantilever array. Chaos **19**, 013138 (2009)
58. M. Kimura, T. Hikihara, Experimental manipulation of intrinsic localized modes in macromechanical system. Nonlinear Theory Its Appl. IEICE **3**, 233–245 (2012)
59. R. Morandotti, U. Peschel, J.S. Aitchison, H.S. Eisenberg, Y. Silberberg, Dynamics of discrete solitons in optical waveguide arrays. Phys. Rev. Lett. **83**, 2726–2729 (1999)

60. J.W. Fleischer, M. Segev, N.K. Efremidis, D.N. Christodoulides, Observation of two-dimensional discrete solitons in optically induced nonlinear photonic lattices. Nature **422**, 147–150 (2003)
61. D.N. Christodoulides, F. Lederer, Y. Silberberg, Discretizing light behaviour in linear and nonlinear waveguide lattices. Nature **424**, 817–823 (2003)
62. A.V. Gorbach, S. Denisov, S. Flach, Optical ratchets with discrete cavity solitons. Opt. Lett. **31**, 1702–1704 (2006)
63. O.A. Egorov, F. Lederer, Spontaneously walking discrete cavity solitons. Opt. Lett. **38**, 1010–1012 (2013)

Chapter 5
Nano-optomechanics by Tailored Light Fields Under Fluctuations

Takuya Iida, Syoji Ito, Shiho Tokonami and Chie Kojima

Abstract We have developed guiding principles to control the dynamics and functions of nanocomposites by optically modulating the balance between the inter-object interaction and the thermal fluctuations. In this chapter, we will show our recent achievements on the development of new theoretical methods to explore the dynamics of nanoparticles (NPs) by the light-induced force (LIF) of tailored light fields under thermal fluctuations. The highly precise optical screening of NPs with the help of fluctuations has been proposed being inspired by a biomolecular motor. Furthermore, the spatial configuration and the collective phenomena of metallic NPs can be simultaneously controlled by LIF and fluctuations. We will explain an experimental demonstration of the selective optical assembling of uniform metallic NPs by LIF of axially-symmetric polarized beams. Medical applications and biosensor applications of assembled metallic NPs are also discussed. Our achievements will pioneer a new research field "Biomimetic Optical Manipulation" based on the fluctuation-mediated nano-optomechanics.

T. Iida (✉)
Graduate School of Science, Osaka Prefecture University, 1-1 Gakuencho,
Nakaku, Sakai, Osaka 599-8531, Japan
e-mail: t-iida@p.s.osakafu-u.ac.jp

S. Ito
Graduate School of Engineering Science, Osaka University, 1-3 Machikaneyama-cho,
Toyonaka, Osaka 560-8531, Japan
e-mail: sito@chem.es.osaka-u.ac.jp

S. Tokonami
Nanoscience and Nanotechnology Research Center, Osaka Prefecture University,
1-2 Gakuencho, Nakaku, Sakai, Osaka 599-8570, Japan
e-mail: s-tokonami@21c.osakafu-u.ac.jp

C. Kojima
Graduate School of Engineering, Osaka Prefecture University, 1-1 Gakuencho,
Nakaku, Sakai, Osaka 599-8531, Japan
e-mail: kojima@chem.osakafu-u.ac.jp

© Springer International Publishing Switzerland 2015
M. Ohtsu and T. Yatsui (eds.), *Progress in Nanophotonics 3*,
Nano-Optics and Nanophotonics, DOI 10.1007/978-3-319-11602-0_5

5.1 Introduction

Conventionally, thermal fluctuations inducing Brownian motion were considered as disturbing factors in the mechanical manipulation of nano-objects. However, paying attention to the behavior of biological nanosystems, they utilize "thermal fluctuations" arising from the collisions of medium molecules to realize the various excellent functions. Assembly processes with external perturbations under fluctuations and dissipation [1] have produced such a variety of biological and non-biological nanosystems [2–4]. For example, a molecular motor like myosin transports materials with the help of fluctuations by changing the symmetry of potential profiles on a one-dimensional helical fiber (actin filament) [5]. Also, through the selective evolution process under the sunlight and the environmental fluctuations, a photosynthetic bacterium obtained a two-dimensional light harvesting antenna (LHA) [6, 7] consisting of ring-like arrangement of interacting dye-molecules that has an optical absorption band optimized for the collection of sunlight at its territory [8]. A three-dimensional bubble-like structure (vesicle) in the biological cells contains nanomaterials necessary for life, and flexibly changes its surface membrane for the metabolism near room temperature by controlling the force between constituent organic molecules with exchange of chemical substances [9]. The common point is that these biological systems exploit the energy of random thermal fluctuations for the transport, assembling phenomena, modulation of spatial configurations through a spatiotemporal modulation of interaction potential by external stimuli.

Turning our eyes to previous works on optomechanics with light-induced force (LIF), this force has allowed us to manipulate and trap micronparticles [10], metallic and semiconductor nanoparticles (NPs) [11–13]. In addition, LIF can be used for the control of the inter-object interaction in the modulation of physical process and chemical reactions [14, 15]. Also, there are reports on the optical manipulation using specialized light fields, for example, the trapping of multiple particles by holographic tweezers [16] or femtosecond pulses [17], and the trapping of molecules by localized field in a metallic nanogap [18]. These features of LIF can be used for controlling the relation between the environmental fluctuations and the interaction potential of nanostructures in a optical manner.

Learning from these unique features of biological systems and optomechanics, we have tried to pioneer guiding principles to control the nonequilibrium dynamics and functions of nanocomposites by optically modulating the balance between the inter-object interaction and the thermal fluctuations leading to "Biomimetic optical manipulation" (Fig. 5.1). In this chapter, to construct the foundation of such a new concept, we investigated the dynamics of metallic nanomaterials by LIF of designed light fields under the random collisions of environmental molecules. For this purpose, we developed new theoretical methods to treat the effect of self-consistent LIF and thermal fluctuations, i.e., "Light-induced force nano dynamics method (LNDM)" in time domain [19] and "Light-induced force nano Metropolis method (LNMM)" in energy domain [20]. The degree-of-freedom of nanodynamics can be greatly enhanced since the LIF depends on various properties of excitation light such as

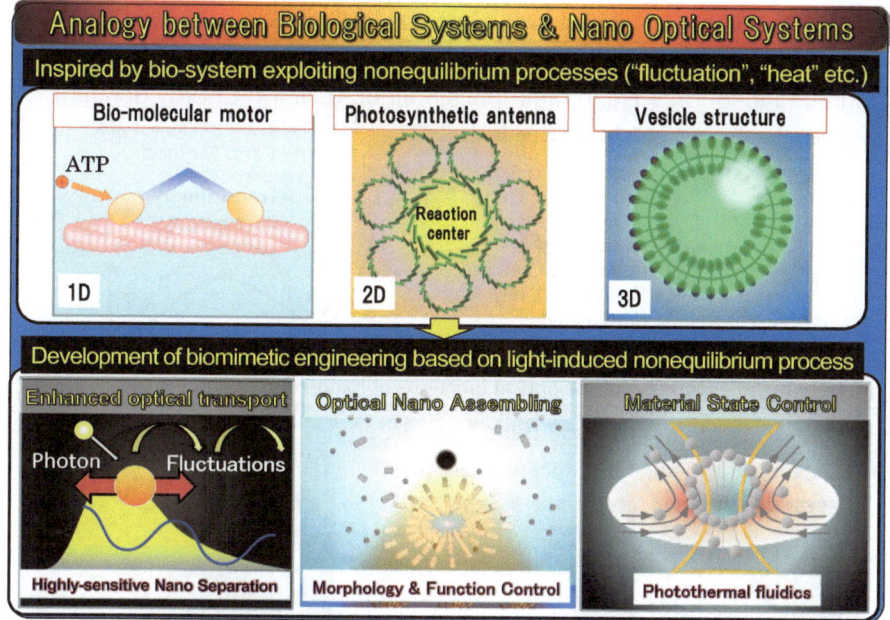

Fig. 5.1 Concept of "Biomimetic Optical Manipulation"

wavelength, angular momentum, polarization and intensity distributions. According to this theory, it has been clarified that efficient optical transport of NPs can be realized by using modulated optical standing wave with the help of fluctuations leading to highly-precise nano-optical separation [21]. Furthermore, we have experimentally demonstrated that ring-like arrangements of metallic NPs with high rotational symmetry similar to natural LHA were selectively created by LIF of doughnut beams under thermal fluctuations [22], which exhibits broadband optical scattering due to the collective phenomena of localized surface plasmons (LSPs). Such collective phenomena of LSPs in densely assembled metallic NPs [23, 24] can be used for the photothermal medical applications [25], and for highly-sensitive optical biosensors based on the coupling of heterogeneous metallic nanostructures [26] and the light-induced bubble to control the local phase transition [27]. Also, the brief explanations of these biological applications will be provided at the final part of this chapter.

5.2 Theory

In this section, we will explain our developed theoretical methods for evaluating time dependence of dynamics of NPs under laser irradiation and thermal fluctuations i.e., LNDM, and for finding the energetically stable and metastable states of

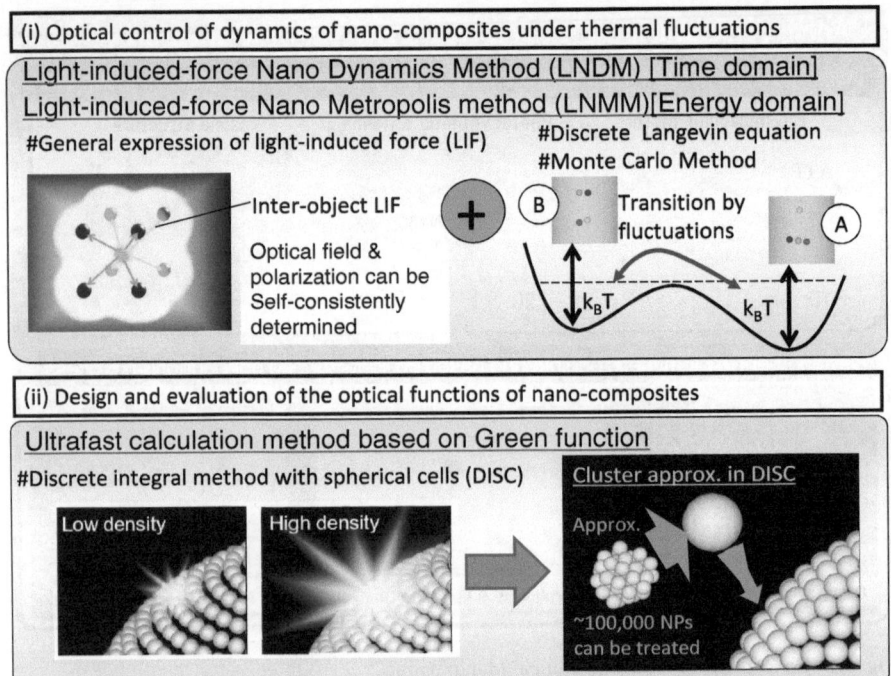

Fig. 5.2 Graphical summary of developed theoretical methods for *light-induced dynamics* of NPs under fluctuations. **a** Optical control of dynamics of nano-composites under thermal fluctuations. **b** Design and evaluation of the optical functions of nano-composites

optically assembled NPs with arbitrary shapes, i.e., LNMM (Fig. 5.2). As the target nanomaterial, we consider metallic NPs exhibiting strong optical response arising from LSP to generate sufficiently strong LIF whose magnitude is comparable to the fluctuation-induced force even at room temperature. Since the response field and the induced electric polarization of collective modes of LSP in lots of densely assembled metallic NPs are necessary for the self-consistent evaluation of LIF. Therefore, we also developed ultrafast calculation method based on the discrete integral method with spherical cells (DISC) that enables us to evaluate the optical response of a vast number of metallic NPs based on the cluster approximation.

5.2.1 Light-Induced-Force Nano Dynamics Method (LNDM)

For the evaluation of the time evolution of mechanical motions of multiple NPs under the LIF and the fluctuation-induced random force, a numerical method called LNDM was developed [19]. This method is based on the general expression of LIF derived from the Lorentz force equation [14] and Brownian dynamics method based on the classical Langevin equation [28]. The Langevin equation can be transformed into a sequential discrete form as follows:

$$\mathbf{r}_i(t + \Delta t) = \mathbf{r}_i(t) + \frac{1}{\xi_i}\mathbf{F}_i^{(EX)}\Delta t + \Delta\mathbf{r}_i^{(R)}, \tag{5.1}$$

where $\mathbf{r}_i$ is the coordinate of the ith NP, $\mathbf{F}_i^{(EX)}$ is an external force including an LIF, and $\xi_i = 3\pi\eta d_i$ is the friction coefficient (d_i is the diameter of the ith NP, and η is the viscosity of the medium). The value of $\Delta\mathbf{r}_i^{(R)}$ is random displacement by a random force $\mathbf{F}_i^{(R)}$ due to thermal fluctuations, and is determined based on the classical Boltzmann distribution. When each time step Δt is much larger than the frictional relaxation time $t_{fric}(= m_i/\xi_i)$ and the displacement during Δt is much smaller than the diameter of each NP, (5.1) can be used, where m_i is the mass of the ith NP. Here, the external force is given using $\mathbf{F}_i^{(EX)} = \mathbf{F}_i^{LIF} + \mathbf{F}_i^{DLVO}$ as a sum of the LIF, $\mathbf{F}_i^{LIF}$, and the spontaneous electrostatic double layer force, i.e., DLVO (Derjagin, Landau, Verwey, and Overbeek) force between NPs in an aqueous solution, $\mathbf{F}_i^{DLVO}$. $\mathbf{F}_i^{DLVO}$ is described as the sum of the attractive van der Waals force and electrostatic repulsive force due to the ionization at the surface of respective NPs [29, 30] (the detail will be explained later). In a previous report of other group, only the LIF and the attractive van der Waals force between metallic NPs were numerically calculated, but the repulsive force was not considered [31–33]. The value of $\mathbf{F}_i^{LIF}$ is evaluated with a general expression of time-averaged LIF [14] as follows

$$\mathbf{F}_i^{LIF} = \frac{1}{2}\text{Re}\left[\sum_\omega \int_V d\mathbf{r}(\nabla\mathbf{E}(\mathbf{r}, \omega)^*) \cdot \mathbf{P}(\mathbf{r}, \omega)\right], \tag{5.2}$$

where $\mathbf{E}$ is the total response electric field as a sum of incident and induced fields, and $\mathbf{P}$ is light-induced polarization. Both a direct LIF from incident light and an inter-object LIF between nano-objects are included. A Drude-type optical susceptibility was used for each metallic NP as follows:

$$\chi_j^{(p)}(\omega) = (\varepsilon_{bg}^{(p)}(\omega) - \varepsilon^{(m)}) - \frac{(\hbar\Omega_{bulk})^2}{(\hbar\omega)^2 + i\hbar\omega(\Gamma_{bulk} + \Gamma_{SD})}, \tag{5.3}$$

where $\varepsilon_{bg}^{(p)}(\omega)$ is the background dielectric function of bulk metal, $\varepsilon^{(m)}$ is the high-frequency dielectric constant of the surrounding medium, Ω_{bulk} is the bulk plasmon resonance frequency, Γ_{bulk} is the bulk non-radiative width, $\Gamma_{SD} = 2V_f/d_i$ is the size-dependent non-radiative damping with $d_i = a_p$ assuming that the diameters of all the NPs are the same, and V_f is the electron velocity at the Fermi level (Fermi velocity). For the evaluation of $\mathbf{F}_i^{LIF}$, $\mathbf{E}$ and $\mathbf{P}$ are determined by self-consistently solving the discrete integral form of a Maxwell equation [34–37] as a linear simultaneous equation of $\mathbf{P}_j$, assuming that the respective NPs are modelled using spherical cells, i.e., discrete integral method with spherical cells (DISC):

$$\mathbf{E}_i = \mathbf{E}_i^{(0)} + \sum_{j=1}^{N} \mathbf{G}_{i,j}^{\mathrm{med}} \mathbf{P}_j V_j, \tag{5.4}$$

$$\mathbf{P}_j = \chi_j \mathbf{E}_j, \tag{5.5}$$

where N is the number of NPs, $V_j = (4\pi/3)(d_j/2)^3$ is the volume of each cell, $\mathbf{E}_i^{(0)}$ is an electric field of incident light, $\mathbf{G}_{i,j}^{\mathrm{med}}$ is Green's function in a homogeneous medium, and χ_j is electric susceptibility in the optical frequency range. We analytically calculated the integral for $i = j$, i.e., the self-term in (5.4), under the assumption that the spatial variation of the internal field is negligible, since we consider NPs whose diameter is much smaller than the light wavelength. While only a simple discrete dipole model is considered here, this treatment can save computational time for an evaluation of NP dynamics. By substituting (5.5) into (5.4), the self-consistent solution of $\mathbf{P}$ is given as

$$\mathbf{P}_j = \sum_{j=1}^{N} \mathbf{A}(i, j)^{-1} \cdot \mathbf{E}_i^{(0)}, \tag{5.6}$$

where $\mathbf{A}(i, j)$ is the coefficient matrix of simultaneous equations of $\mathbf{P}_j$.

The approximated expressions of gradient force and dissipative force (sum of the scattering and the absorbing forces) on a pair of metallic NPs from incident light can be obtained as follows:

$$\langle \mathbf{F}_{\mathrm{grad}} \rangle = \sum_{\omega} \frac{(\hbar \bar{\Omega}_{\mathrm{pl}} + \bar{\Delta}_{\mathrm{int}} - \hbar\omega)\alpha \nabla |\mathbf{E}^{(0)}|^2}{(\hbar \bar{\Omega}_{\mathrm{pl}} + \bar{\Delta}_{\mathrm{int}} - \hbar\omega)^2 + (\bar{\Gamma}^{(\mathrm{Rad})} + \bar{\Gamma}_{\mathrm{int}} + \Gamma^{(\mathrm{NRD})})^2}, \tag{5.7}$$

$$\langle \mathbf{F}_{\mathrm{dis}} \rangle = \sum_{\omega} \frac{(\bar{\Gamma}^{(\mathrm{Rad})} + \bar{\Gamma}_{\mathrm{int}} + \bar{\Gamma}^{(\mathrm{NRD})})\beta |\mathbf{E}^{(0)}|^2}{(\hbar \bar{\Omega}_{\mathrm{pl}} + \bar{\Delta}_{\mathrm{int}} - \hbar\omega)^2 + (\bar{\Gamma}^{(\mathrm{Rad})} + \bar{\Gamma}_{\mathrm{int}} + \bar{\Gamma}^{(\mathrm{NRD})})^2}, \tag{5.8}$$

where $\bar{\Omega}_{\mathrm{pl}}$ is the self-consistently obtained resonance energy of an LSP in a single metallic NP; $\bar{\Gamma}^{(\mathrm{Rad})}$ is the radiative width (radiative relaxation rate) of a single NP; $\bar{\Delta}_{\mathrm{int}}$ and $\bar{\Gamma}_{\mathrm{int}}$ are the modulation of eigenenergy and radiative width as a function of D due to the interaction between NPs, respectively; and α and β are positive coefficients proportional to the laser intensity. $\bar{\Gamma}^{(\mathrm{NRD})} = \Gamma_{\mathrm{bulk}} + \Gamma_{\mathrm{SD}}$ is the sum of nonradiative relaxation rate including the size dependent term. Similar to the case of a semiconductor NP as shown in [14], a dissipative force pushes the NPs toward the direction of light propagation, and a gradient force toward a high intensity region is exerted on NPs when $\hbar\omega$ is lower than the resonance energy of the LSP.

The magnitudes of both components are greatly enhanced near the resonance of the LSP. The inter-object LIF on the ith NP is given by

$$\langle \mathbf{F}_{\text{int}} \rangle_i = \frac{1}{2} \text{Re} \left\{ \sum_\omega \sum_j [\nabla_{ij} \mathbf{G}_{i,j}^{\text{med}} \cdot \mathbf{P}_j V_j]^* \cdot \mathbf{P}_i V_i \right\}, \qquad (5.9)$$

which includes all contributions of multiple scattering from the other NPs. This force behaves attractively and repulsively depending on the light polarization and frequency (wavelength), respectively. Also, in Fig. 5.3, we will show several results by rigorous numerical calculation with (5.2) considering a pair of Au NPs using the following parameters: $\hbar\Omega_{\text{bulk}} = 8.958$ eV; $\Gamma_{\text{bulk}} = 72.3$ meV; $V_f = 0.922$ nm eV; and $\varepsilon_{\text{bg}}^{(p)}(\omega) = 12$ [38, 39] neglecting the interband effect for the essential discussion.

Since the dynamics of metallic NPs in aqueous solution at room temperature is considered here, the spontaneous force between NPs consists of the van der Waals attractive force and the electrostatic repulsive force [29, 30]. The sum of these

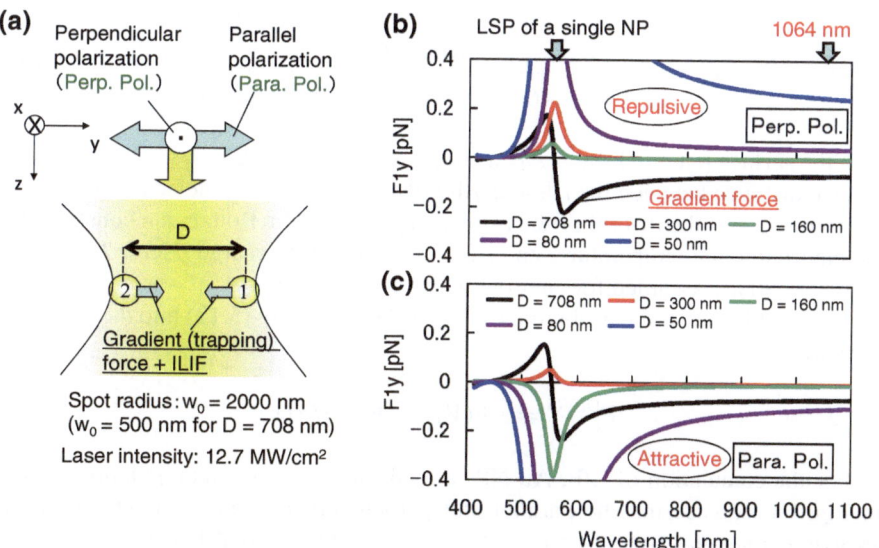

Fig. 5.3 **a** Model for the calculation of LIF on a pair of Au NPs. **b** and **c** Wavelength dependence of gradient (trapping) force and inter-object LIF between Au NPs. The laser intensity is assumed to be 12.7 MW/cm². At $D = 708$ nm, the intensity gradient in the y-direction is at maximum for spot radius of $w_0 = 500$ nm and laser power of 100 mW. For *other lines*, the incident light is almost considered as a plane wave with the same intensity with $w_0 = 2,000$ nm and laser power of 1.6 W. The resonant wavelength of LSP in a single Au NP is given as 555 nm under the assumed condition, the non-resonant light of 1,064 nm wavelength is considered as an incident light for *all lines*. Reprinted (adapted) with permission from [19]. Copyright (2012) American Chemical Society

forces is given as $\mathbf{F}_i^{\text{DLVO}}$ called DLVO force. The attractive van der Waals force is given by

$$V_{\text{vdw}} = -\frac{C_H}{6}\left[\frac{2d_id_j}{4D_{ij}^2 - (d_i + d_j)^2} + \frac{2d_id_j}{4D_{ij}^2 - (d_i - d_j)^2} + \ln\frac{4D_{ij}^2 - (d_i + d_j)^2}{4D_{ij}^2 - (d_i - d_j)^2}\right],$$

(5.10)

where C_H is the Hamaker constant, D_{ij} is the center distance between the ith and jth NPs. In addition, the electrostatic repulsive potential is given by

$$V_{\text{rep}} = 4\pi\varepsilon_{\text{st}}\psi_0^2\frac{d_id_j}{2(d_i + d_j)}\ln[1 + \exp(-\kappa s_{ij})]$$

(5.11)

for $\kappa d_\eta > 10$ ($\eta = i, j$: number of particles), and

$$V_{\text{rep}} = 4\pi\varepsilon_{\text{st}}\frac{d_i}{2}\frac{d_j}{2}Y_iY_j\left(\frac{k_BT}{e_0}\right)^2\frac{\exp(-\kappa s_{ij})}{D_{ij}}$$

(5.12)

for $\kappa d_\eta < 10$, where we have

$$Y_\alpha = \frac{8\tanh(\Psi_0)}{1 + [1 - 4(\kappa d_\eta + 1)/(\kappa d_\eta + 2)^2\tanh(\Psi_0)]^{1/2}}$$

(5.13)

with the surface potential $\Psi_0 = \varepsilon_{\text{st}}\psi_0/4k_BT$, $s_{ij} = D_{ij} - (d_i + d_j)/2$ is the shortest surface distance between the ith and jth NPs, ε_{st} is the static dielectric constant of the medium, κ is the inverse of the Debye length, k_B is the Boltzmann constant, T is the environmental temperature, and e_0 is the elementary charge (Here, $T = 298\,\text{K}$ and $\varepsilon_{\text{st}} = 80$ are assumed for environmental parameters).

The total DLVO force is obtained by substituting (5.10)–(5.13) into the following equation:

$$\mathbf{F}_i^{\text{DLVO}} = -\partial(V_{\text{rep}} + V_{\text{vdw}})/\partial D_{ij}$$

(5.14)

For the calculation of LIF, Au NPs are considered as the manipulation targets. In Fig. 5.3, considering the parallel and perpendicular (y and z) polarizations of an axis connecting two Au NPs, I have calculated the total LIF in the x and y directions for different distances D between the two Au NPs. First, we numerically calculated the LIF for the two Au NPs at the configuration with a maximum intensity gradient in the y-direction (curves for $D = 708\,\text{nm}$ in Fig. 5.3b, c). Under this configuration, the interaction between the two Au NPs is very weak, and thus, the LIF on each NP shows properties similar to those of an LIF on a single NP. Moreover, the maximum value of gradient force is estimated as of the order of several tens of femtonewtons. This value corresponds to the force experimentally estimated in [40]. Particularly, when D is small, an inter-object LIF can be comparable to a direct

LIF by incident light. An attractive LIF arises between the Au NPs for parallel polarization, and the LSP resonance peak greatly moves toward the long wavelength region corresponding to the electromagnetic bonding state (Fig. 5.3b). Inversely, for perpendicular polarization, a repulsive LIF arises, and the peak of LSP resonance slightly moves toward the short wavelength region corresponding to the antibonding state (Fig. 5.3c).

Next, Fig. 5.4 shows the D-dependence of DLVO potential and the sum of LIF and DLVO forces between two Au NPs, where (5.7)–(5.10) and (5.2)–(5.5) were used. To investigate the essential properties of the dynamics, it is assumed that parameters for the DLVO potential provide a steep and strong repulsive force, where the surface potential is $\psi_0 = -90$ mV, the Debye length $1/\kappa = 6.08$ nm, and the Hamaker constant $C_H = 1.602 \times 10^{-19}$ J [19]. The height of the assumed potential barrier is of the order of several tens of $k_B T$ (Fig. 5.4a). The surface potential can be controlled by changing the protecting agent or by adding electrolytes [29, 30]. For y-polarization, as the light intensity increases, D for a stable point decreases, and the potential well deepens by the attractive inter-object LIF (Fig. 5.4b). On the other hand, for z-polarization, the potential well generated by a focused laser beam deepens, but D for a stable point slightly increases by the repulsive inter-object LIF (Fig. 5.4c). In the

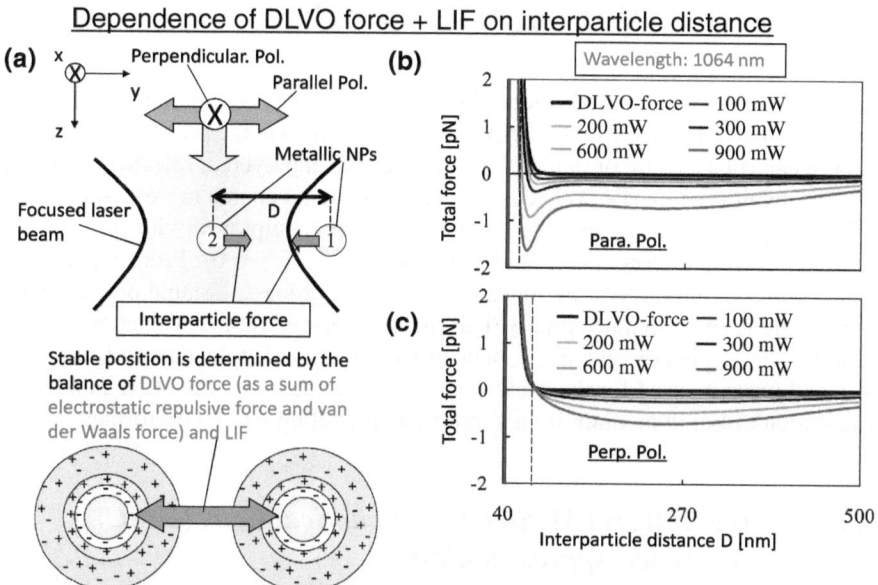

Fig. 5.4 (**a**) (*Upper*) Model for the calculation of distance dependence of interparticle force between metallic NPs. (*Lower*) Schematic illustration of interparticle force including electrostatic force due to the ionization of surface of metallic NPs and LIF. (**b**) and (**c**) Dependence of total force, as a sum of LIF and DLVO potential, on center distance between two Au NPs for various laser intensities under y- and z-polarization, respectively. The *vertical axis range* is limited from -2 to 2 pN to facilitate visualization. Reprinted (adapted) with permission from [19]. Copyright (2012) American Chemical Society

calculation results presented herein, two Au NPs can be aligned parallel to the light polarization, where the potential minimum is generated through a balance of strong attractive inter-object LIF and DLVO force parallel to the incident light polarization. The estimated D for this stable state is approximately 56 nm for 900 MW laser intensity at a parallel configuration (Fig. 5.4a).

5.2.2 Light-Induced-Force Nano Metropolis Method (LNMM)

Next, in order to find an energetically stable (or metastable) state of multiple NPs under the LIF and the thermal fluctuations, a theoretical method called LNMM was also developed [20, 22]. Under the assumed condition, NPs are affected by LIF and show the stochastic motion due to the fluctuations in liquid medium. Metastable states can be found after random spatial displacement of NPs based on following Metropolis algorithm. At each step, NPs are randomly moved according to the next transition probability,

$$p = \min \left[1, \exp\left(-\Delta H / k_B T\right) \right] \tag{5.15}$$

$$\Delta H = -\sum_{i}^{N} \mathbf{F}_i^{(EX)} \cdot \Delta \mathbf{r}_i + \Delta V^{DLVO}, \tag{5.16}$$

where ΔH is the variation of potential in the total system before and after the random motion given by the LIF and DLVO theory as shown in the previous Sect. 5.2.1. (5.15) means that a result of random displacement is employed or rejected according to the probability of $\exp\left(-\Delta H / k_B T\right)$ when the system become more unstable with higher energy ($\Delta H > 0$), and that the displacement is employed with $p = 1$ when the system become more stable with lower energy ($\Delta H < 0$). From a particular initial configuration of NPs, by repeating the above process, a stable or metastable configuration can be obtained finally. This method is useful for the evaluation of light-induced arrangement of anisotropic nanostructures (for example, nanorod, nanotube etc.), and it can be used for the evaluation of arrangement process of NPs under the evaporation of liquid medium if an appropriate modeling will be performed.

5.2.3 Discrete Integral Method with Spherical Cells (DISC) Under Cluster Approximation

The collective phenomena of LSPs in densely assembled metallic NPs is significantly important for the understanding of dynamics of light-induced dynamics and a variety of optical applications. In order to evaluate the collective dynamics of LSPs in such a system, it takes very long time with usual DISC if the number of NPs is very large. For example, a chemical self-assembled method enables us to fix several hundreds of thousands of metallic (Au or Ag) NPs onto a micron plastic bead [24, 41], which is called a metallic nanoparticle-fixed bead (MNFB) as discussed in the later Sect. 5.4. Therefore, the effective theoretical method was required to treat the optical response of such a MNFB. Therefore, the DISC with the cluster approximation, i.e., "cluster DISC method" was proposed [26]. For the numerical evaluation of optical response of high density metallic NP system, it is assumed that a cluster consists of several tens of NPs as a discrete cell. Under this approximation, the integral form of Maxwell equation, i.e., (5.4)–(5.6) can be slightly modified as follows:

$$\mathbf{E}_i = \mathbf{E}_i^{(\text{inc})} + \sum_{j \neq i}^{N^{(c)}} \mathbf{G}^{(\text{m})}(\mathbf{r}_{ij}) \mathbf{P}_j V^{(c)} + \mathbf{S}_i \mathbf{P}_i \tag{5.17}$$

$$\mathbf{P}_j = \chi_j^{(\text{cp})}{}_j \mathbf{E}_j, \tag{5.18}$$

where $N^{(c)} = N^{(\text{T})}/N^{(\text{CNP})}$ is the number of clusters, $N^{(\text{T})}$ is the total number of NPs, $N^{(\text{CNP})} = V^{(c)}/\{(4\pi/3)(a_p/2)^3\}$ is the number of NPs in each cluster, $V^{(c)} = (4\pi/3)(D_c/2)^3$ is the volume of each cluster, D_c is the diameter of each cluster, $\mathbf{E}_i^{(\text{inc})}$ is the incident electric field, $\mathbf{G}_{i,j}^{(\text{m})}$ is a Green's function in a homogeneous medium, and $\chi_j^{(\text{cp})}$ is the optical susceptibility of each NP contained in a cluster, where the diameter is set to $d_i = a_p$ in the size-dependent nonradiative damping Γ_{SD} of (5.3) rather than D_c. The integral $\mathbf{S}_i = \int_{V^{(c)}} d\mathbf{r}' \mathbf{G}^{(\text{m})}(\mathbf{r}_i - \mathbf{r}')$ in (5.17) means the self-term for $i = j$ in each cluster assuming that the spatial variation of the internal field is very small, which is valid when D_c is sufficiently smaller than the wavelength of the incident light. In Fig. 5.5, an example of calculations with "cluster DISC method"

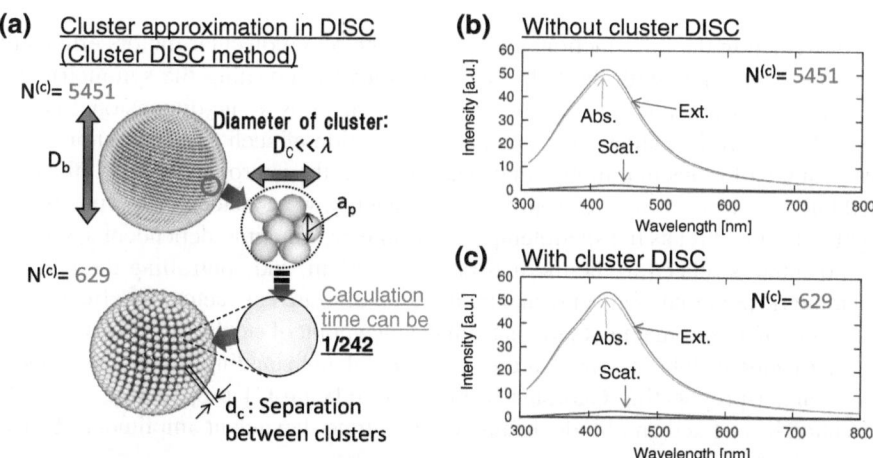

Fig. 5.5 **a** Schematic image of model of a AgNP-fixed bead (AgNP-FB) for the calculation under the cluster approximation. **b** Calculated spectra of extinction, absorption, scattering of AgNP-FB without cluster approximation. **c** Calculated spectra of extinction, absorption, scattering of AgNP-FB with cluster approximation. Reprinted (adapted) with permission from [26]. Copyright (2014) American Chemical Society

of optical response of densely assembled Ag NPs fixed on a transparent sphere like a organic polymer bead with low refractive index is shown, where the dielectric constant near that of water. Also, for Ag NPs on a bead, we use the following values: $\hbar\Omega_{bulk} = 9.088$ eV; $\Gamma_{bulk} = 21.23$ meV; $V_f = 0.922$ nm eV, and $\varepsilon_{bg}^{(p)}(\omega) = 5.0$ neglecting the interband effects beyond the observed wavelength region. The extinction spectrum is proportional to the dissipative force by a propagating plane wave [14, 42] as follows $< \mathbf{F}_{total}^{Ext} > = < \mathbf{F}_{total}^{Scat} > + < \mathbf{F}_{total}^{Abs} >$, where $< \mathbf{F}_{total}^{Scat} >$ is the component proportional to the scattering and $< \mathbf{F}_{total}^{Abs} >$ is the absorption component proportional to the absorption. The extinction, the scattering, and the absorption of NPs can be evaluated by using this relation, respectively. We can see very good agreement between the results with and without the cluster DISC method. Moreover, the calculation time can be reduced into 1/242. Such a speed up of calculation is more prominent when the number of NPs is very large. For example, when the number of Ag NPs is 200,000 and they are approximated with 3,462 cells, the calculation time can be 1/3,000 and we can confirm a good agreement with experimental result [26]. Even with a recent CPU in the workstation, the calculation of optical response of 200,000 NPs takes about 8 years without the cluster approximation. However, we can perform the almost equivalent calculation only within 1 day by using our developed cluster DISC method.

5.3 Biomimetic Optical Manipulation Under Thermal Fluctuations

5.3.1 Fluctuation-Mediated Nano Optical Screening

As described in the introduction, a myosin as one of a molecular motors efficiently transports materials with the help of fluctuations by repeating the symmetric and asymmetric of potential profiles on an actin filament as a one-dimensional helical fiber [5]. Figure 5.6a shows the schematic illustration of such a phenomenon. From the analogy of molecular motor, in order to convert the isotropic energy of the thermal fluctuations into an anisotropic transport energy, we consider a superposition of light fields that breaks the spatiotemporal symmetry. The time-dependent asymmetric potential is related to the thermal ratchet problem, and controlling the periodic asymmetric potential well depth and modulation time are key factors for effective NP transport under thermal fluctuations from the viewpoint of stochastic resonance [43].

The incident light is considered to be a modulated standing wave consisting of two counter-propagating Gaussian beams. While beam G1 propagating in the +z direction has a fixed amplitude, beam G2 has a time-dependent amplitude. The two beams are described as

$$\mathbf{E}_{G1} = \hat{\mathbf{y}} E_0 u_g \exp[ikz] e^{-i\omega t} \tag{5.19}$$

$$\mathbf{E}_{G2} = \hat{\mathbf{y}} E_0 u_g^* f(\Omega t) \exp[-ikz] e^{-i\omega t}, \tag{5.20}$$

Optical selection of NPs inspired by bio-molecular motor

(a) Molecular motor for transport with the help of fluctuations

(b) Transport by spatiotemporally modulated optical standing wave

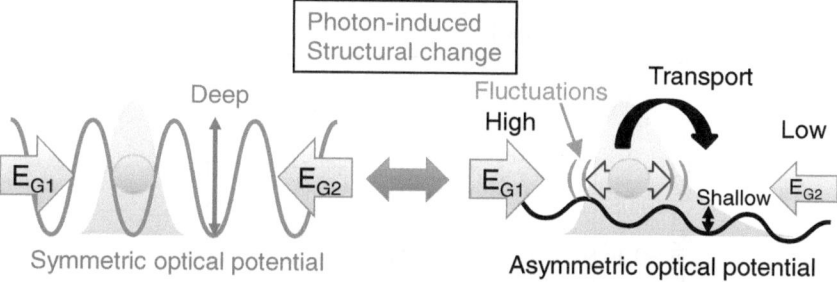

Fig. 5.6 Analogy between bio-molecular motor and transport of NPs by spatiotemporally modulated optical standing wave. Optical selection of NPs inspired by bio-molecular motor. **a** Molecular motor for transport with the help of fluctuations. **b** transport by spatiotemporally modulated optical standing wave

where $\hat{y}$ is the unit vector in y-direction, E_0 is the electric field intensity at the center of the spot. The wave number in the medium is $k = n(\omega/c)$, n is the refractive index of surrounding medium, $u_g = -iw_0^2 Q(z) \exp\left[iQ(z)(x^2 + y^2)\right]$ is the normalized electric field distribution of the Gaussian beam with $Q(z) = k/(2z - ikw_0^2)$, and w_0 is the spot radius, where it is assumed to be $w_0 = 1.0\,\mu$m in the numerical calculations. Laser wavelength is assumed to be 780 nm can be generated with a Ti: Sapphire laser source. While the maximum power of G1 and G2 are the same, the amplitude of G2 changes according to the amplitude modulation function as follows

$$f(\Omega t) = b + (1 - b)\frac{\cos\left[\Omega t\right] + 1}{2}, \ (0 \le b \le 1), \qquad (5.21)$$

where b is the amplitude modulation parameter. Such a modulation can be easily realized if an attenuating filter is inserted into the light path and rotates periodically with a slow angular frequency Ω. This $f(\Omega t)$ is called amplitude modulation function, which varies periodically as $\cos\left[\Omega t\right]$; when $\cos\left[\Omega t\right]$ is equal to 1, $f(\Omega t)$ is also equal to 1, and the amplitude of G2 takes its maximum value, giving a standing-wave intensity distribution as shown in Fig. 5.7a. In contrast, when $\cos\left[\Omega t\right]$ is equal to -1, then $f(\Omega t)$ is equal to b, and the maximum intensity of the standing wave decreases

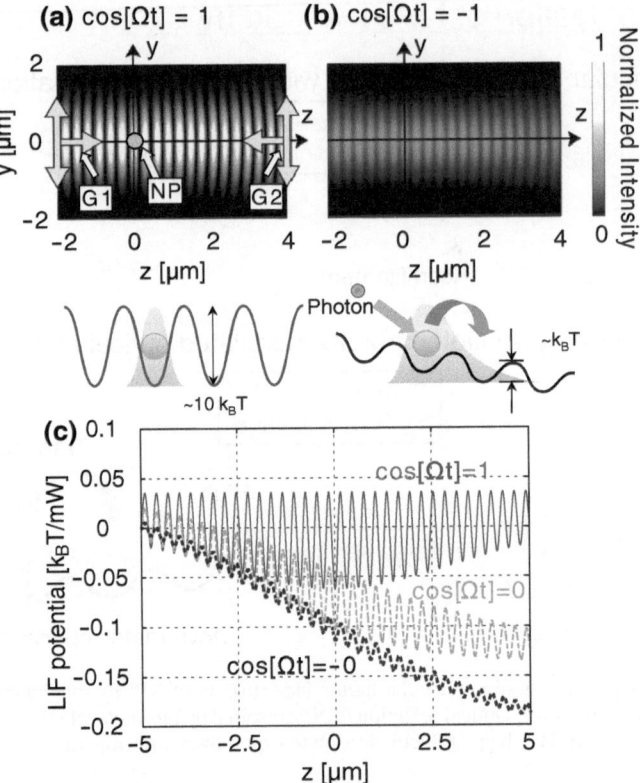

Fig. 5.7 **a** and **b** Spatial distributions of electric field intensity of the temporally modulated standing wave consisting of two Gaussian beams (wavelength: 780 nm) at $\cos[\Omega t] = 1$ and $\cos[\Omega t] = -1$, respectively. Amplitude modulation parameter is set to b = 0.20 and Both distributions are normalized by the maximum value in (**a**). A single Au NP of 40 nm in diameter is initially located at the origin in a region of high intensity. **c** The z-position dependence of the optical potential that the NP experiences in the spatiotemporally modulated standing wave. Reprinted (adapted) with permission from [21]. Copyright (2012) American Chemical Society

as shown in Fig. 5.7b. The depth of the optical potential well for a laser of several hundred milliwatts is comparable to the energy of the thermal fluctuations at room temperature under the assumed condition for the optical manipulation of metallic NPs. For example, the considered laser intensity of each beam is 12.73 MW/cm² for 400 mW input power and $w_0 = 1.0\,\mu$m. If the variation of the trapping potential created by the standing wave is tuned to the diffusion velocity of the thermal fluctuations, NPs can escape the optical potential wells with the help of thermal fluctuations. In the present case, it is assumed that the standing wave vibrates slowly with a timescale comparable to the hopping between several optical potential wells by diffusion, and we employ a time cycle of $\Omega = 8\pi$ [rad/s] corresponding to a vibration of 4 times per 1 s to utilize the fluctuations effectively. The diffusion length is approximately estimated as 1.6 μm during 0.1 s. The laser wavelength is tuned to 780 nm giving optimum dissipative and gradient force magnitudes for Au NPs in a medium of water at room temperature ($T = 298$ K gives a thermal energy of $k_B T = 26$ meV as the criteria for the optical potential, where k_B is the Boltzmann constant). In the assumed condition, the balance of dissipative and gradient forces on a NP changes according to the intensity and spatial variation of the standing wave (Fig. 5.7c). Within the time range that $\cos[\Omega t] < 1$ is satisfied, the negative slope of the potential well envelope function originates from the dissipative force since the photon momentum from beam G1 exceeds that of beam G2. Similarly to the case of molecular motor as shown in Fig. 5.6, periodic potential wells generated by the standing wave show transiently an asymmetric spatial structure as the amplitude of G2 is decreased.

In order to confirm the effects of the fluctuations, the dynamics of each NP was evaluated in the absence of thermal fluctuations. At the initial state, NPs are trapped by the gradient force at the origin where the optical field has a high intensity, and this gradient force gradually weakens as electric field intensity of G2 decreases and potential wells gets shallower. Meanwhile, the dissipative force on the NPs increases and pushes NPs in the propagation direction of G1 as the intensity of G2 is decreased. Since the intensity of G2 always has a finite value, the NPs cannot escape the initial potential well without the help of thermal fluctuations. In the realistic system, the effect of the thermal fluctuations induces Brownian motion of NPs arising from the random collisions of the medium molecules. For the maximum amplitude of the standing wave, the gradient force is dominant and random force from the thermal fluctuations does not affect on the trapping of NPs with a sufficiently high laser intensity. For example, a potential well depth can be greater than $10\,k_B T$ for 400 mW input laser power. When the intensity of G2 decreases and the gradient force weakens, the effect of thermal fluctuations becomes non-negligible. In this case, NPs are pushed into positive direction by the dissipative force arising from G1, and the thermal fluctuations accelerate the NP hopping between neighboring potential wells of depths less than $10\,k_B T$ even for a laser power of 400 mW. The trapping stiffness under the balance of gradient and dissipative forces changes depending on the size of the NP, larger NPs are trapped more strongly and stay near the initial potential well due to the relatively weak fluctuation effect. On the other hand, the gradient force on smaller NPs is weaker and the effect of fluctuations becomes relatively stronger. In detail, by controlling the balance of optical trapping, optical transport, and thermal fluctuation, we can selectively extract NPs with particular sizes under an optimized balance of these three factors. Also, it has been confirmed that the transport distance can be enhanced by the effect of fluctuations even when many interacting NPs with the interval of several hundred nanometers are in the laser spot as an initial condition. The dissipative force can be also enhanced due to the red-shift of LSP resonance arising from the electromagnetic interaction as discussed in 5.3.2 later. Such a spectral modulation of LIF could be useful for the efficient transport by infrared light. This possibility was examined in detail, where results of the LNDM simulation based on (5.1) were given for Au NPs in a modulated standing wave defined by (5.19) and (5.20). In Fig. 5.8, an example of calculated time dependence of dynamics of 40 nm NP is shown. Since the NPs exhibit hopping between neighboring potential wells during the short times when the sum of the dissipative and random

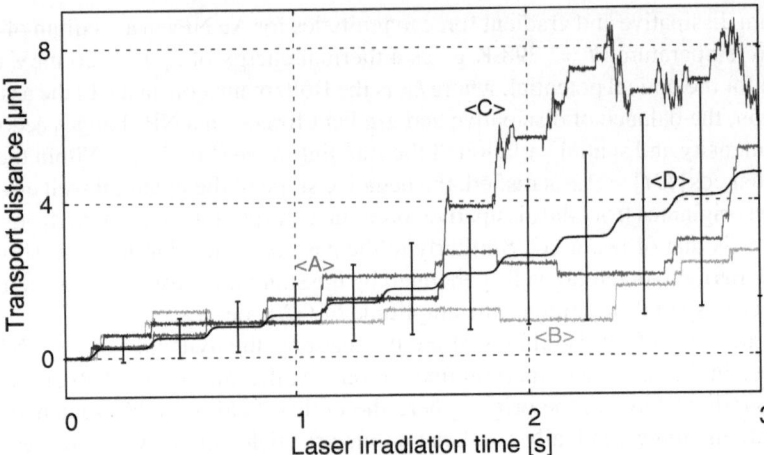

Fig. 5.8 Transport distance of a single Au NP of 40 nm in diameter as a function of time during 3 s (15,000,000 steps) with 300 mW and b = 0.20. Positions of respective NPs were recorded every 0.05 s (250,000 steps). The simulations were repeated 2,000 times. Three of simulation results <A>, <B> and <C> were plotted as examples. Line <D> indicates the statistical average of the time dependence of the z-displacement was plotted. *Vertical bars indicate* the standard deviation

forces exceeds the gradient force as the amplitude of G2 becomes lower, the averages of the z-displacement also show discrete step-like behavior.

Figure 5.9a reveals the results for various diameters of NPs (40, 50, and 60 nm) and the possibility of separating intermediate-sized NPs (50 nm diameter). To control the separation by setting the amplitude modulation parameter $b = 0.20$ and tuning laser power into 200 mW. For a large b, the degree of asymmetry decreases due to the modulation of the G2 amplitude and the minimum gradient force on the NPs increases. On the other hand, for a small b, the amplitude modulation of G2 is larger and the minimum gradient force is weaker. In addition, while the NPs of 40 nm in diameter are widely dispersed and become extremely dilute due to the strong diffusion from the thermal fluctuations, the NPs of 50 nm in diameter show unidirectional transport and was concentrated at a distance less than 4 μm from the origin. Meanwhile, the NPs of 60 nm in diameter remain trapped in the initial potential well. Since a difference in b is related to the spatiotemporal symmetry of the standing wave potential, these results indicate that we can extract particular NPs of intermediate sizes by modulating the symmetry of the light field. While the averaged transport distance seems to be less than 10 μm during 3 s for the assumed condition, it can be controlled by changing w_0 and laser irradiation time. The decay of the laser intensity in z-direction can be suppressed for large laser spot, which leads to long transport distance. This tendency can be used for the easy observation of dynamics with optical microscope. Moreover, in Fig. 5.9b, we consider NPs of 40 nm in diameter as the separation targets by using high power laser of 400 mW. Larger NPs of 50 and 60 nm in diameter were strongly trapped at the starting point, whereas the small

Fig. 5.9 **a** and **b** Dynamics of NPs with three types diameters (40, 50, and 60 nm) in modulated standing wave of different laser powers of 200 and 400 mW. **c** Time averaged transport distance as a function of laser power for NPs of respective sizes, which can be interpreted as the dependence on the magnitude of fluctuations relative to laser power

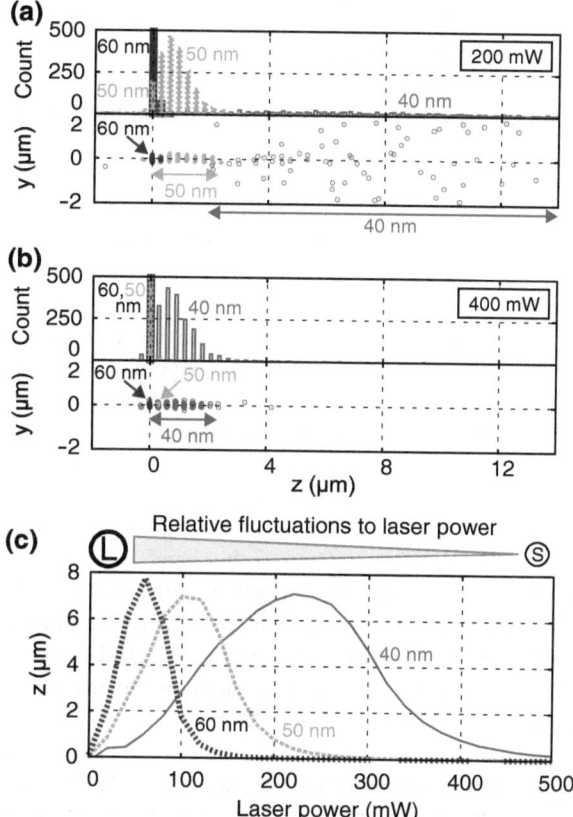

NPs of 40 nm in diameter were efficiently transported and separated. The statistical averages of the NP displacement during 3 s are evaluated as functions of laser power for NPs of respective diameters (Fig. 5.9c). The peak structure at a particular laser power appears and its position is different for each diameter. The effect of fluctuations can be controlled by laser power, and the resonance structure for particular fluctuation intensity is similar to the "Stochastic resonance" [43] as one of various nonequilibrium phenomena.

The results and discussion here explain the mechanism of efficient unidirectional transport of the extracted NPs can be realized by adjusting the intensity and variation of the standing wave amplitude under thermal fluctuations. Particularly, we have clarified the extracting NPs with diameters on the order of 10 nm, we can realize selective manipulation with a higher accuracy by optimizing conditions such as the viscosity and temperature of the media, and the wavelength and polarization of the incident light. This principle can be called "Fluctuation-mediated Optical Screening (FMOS)", which will open the way to a next-generation optical nano-separation at room temperature.

5.3.2 Control of Dynamics and Optical Properties of Nanoparticles

In this subsection, we will discuss the possibility of the simultaneous control of configuration and optical properties of metallic NPs [19] before the detailed discussion on the analogy of the evolution processes of natural LHA as an example of two dimensional optical manipulation in the next Sect. 5.3.3. First, the optical arrangement of Au NPs in aqueous suspension with linearly-polarized light is considered. When the intensity of focused laser increases, the diffusion of Au NPs arising from a random collision of water can be suppressed and a pair of Au NPs can be stably trapped. The relation between LIF and the laser intensity (100–900 mW) is consistent with that in previous experimental papers [40, 44]. The photo-induced heat may be concerned, whereas it would be at most 10 % viscosity changes within the considered intensity region and for NP of 40 nm diameter from [45]. This can be estimated from the relation that heat is proportional to the volume of NP [23]. In order to reduce the photothermal effect heavy water D_2O can be used [46].

In Fig. 5.10b, c, we can see that a small number of Au NPs can be trapped parallel to the linear polarization. On the other hand, when the number of Au NPs (N) increases, some of the Au NPs are trapped at a different z-position from their first linear chain due to the limitation of the width of optical potential well in the y-direction depending on the diameter of the laser spot (Fig. 5.10d). In this case, the positions of the Au NPs in the axial direction are actually determined by the interaction in the laser propagation direction along z-axis.

In Fig. 5.11, we have investigated the optical spectra (extinction, scattering and absorption) depending on the number of Au NPs assembled by linearly polarized focused beam. From Fig. 5.11a–c, it is confirmed that total scattering spectra exhibit the enhancement of the peak value and width with increasing N. Also, in the normalized scattering spectra, the peak position moves toward a longer wavelength region, and the peak width becomes larger as N increases. As indicated by in (5.7) and (5.8), the nonradiative decay rate, $\bar{\Gamma}^{(NRD)}$, is independent of N, which implies an increase in the radiative decay rate. The N-dependence of the radiative decay rate is also illustrated in Fig. 5.11e, which is obtained by subtracting the nonradiative width ($\bar{\Gamma}^{(NRD)} = 118.4$ meV for 40 nm Au NPs assumed here) from the total peak width using eV as the unit of the horizontal axis (proportional to the inverse of the wavelength). The value of the radiative relaxation rate, which is on the order of hundreds meV and is comparable to $\bar{\Gamma}^{(NRD)}$, gives an ultrafast optical response during an inverse of several tens of femtoseconds. This phenomenon is similar to superradiance, which has been previously reported in the case of atoms and quantum dots [47–51]. The main origin is due to the enhanced dipole moment arising from the collective interaction of excited states in many particles via light electromagnetic field. Since the induced polarization of excited LSPs leads to the radiative relaxation rate enhancement here, this effect can be called "Plasmonic superradiance". $\bar{\Gamma}_{int}$ in (5.7) and (5.8) in 5.2.1 plays an important role in such an enhancement. Since scattering cross section is proportional to the radiative decay rate divided by the resonant denominator, and absorption cross section is proportional to the nonradiative decay

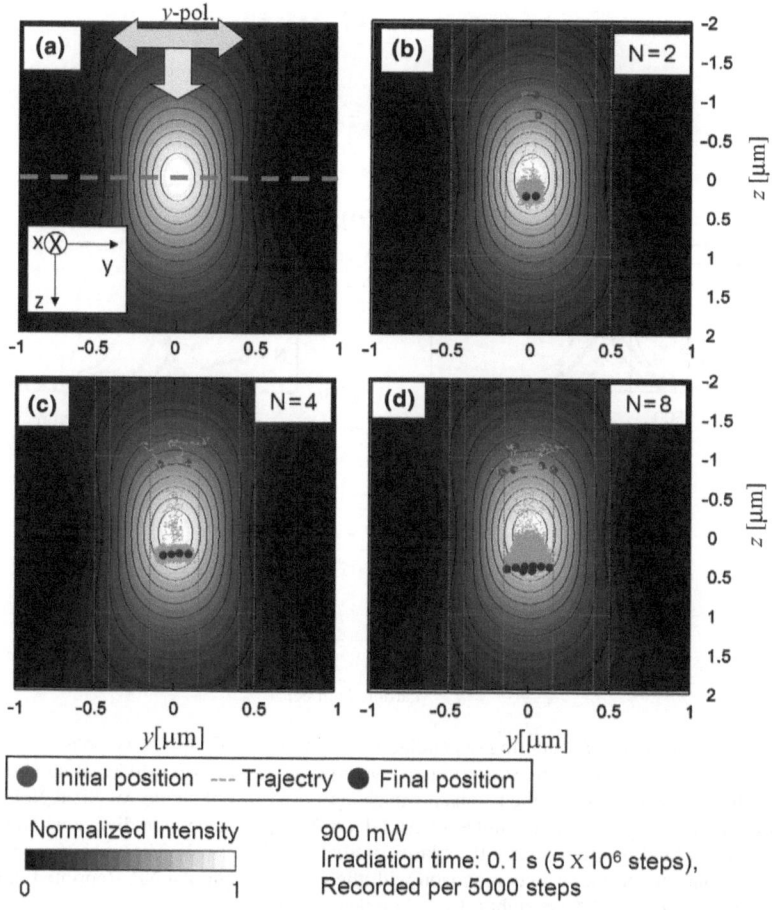

Fig. 5.10 Calculated dynamics of Au NPs under the irradiation of focused laser and the thermal fluctuations. **a** Distribution of laser intensity near the focal point as density and contour plot. The broken horizontal line contains the beam waist. **b–d** Dynamics of different number of Au NPs under stable trapping with high laser power. The spot diameter is set to 1 μm. Reprinted (adapted) with permission from [19]. Copyright (2012) American Chemical Society

rate divided by the resonant denominator [14], strong light scattering can be obtained with a large radiative width. This indicates that we can realize simultaneous control of the spatial configuration and the optical properties of metallic NP-assembly by LIF and thermal fluctuations.

Moreover, paying attention to an axially-symmetric polarized vector beam [52, 53], we have theoretically revealed that Au NPs can be assembled into a ring-like structure similarly to a LHA in a photosynthetic bacterium (Fig. 5.12) [54]. Particularly, Au NPs can be densely assembled due to the attractive force parallel to the azimuthal polarization of an assumed doughnut beam (Fig. 5.12b). Also, large spectral broadening and red-shift can be confirmed, and these spectral modulation can

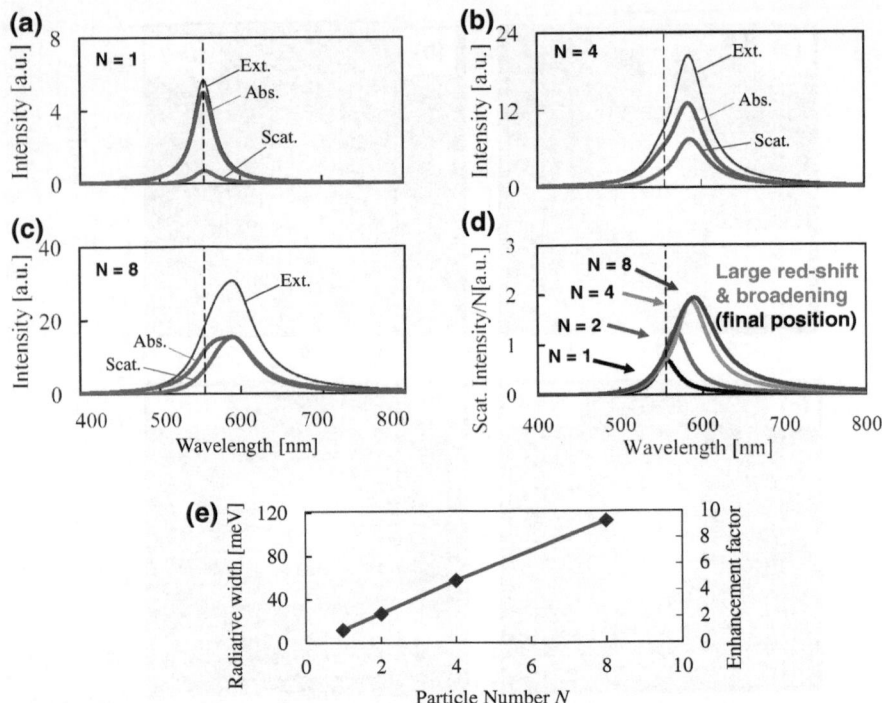

Fig. 5.11 Enhancement of scattering depending on the number of Au NPs (y-polarized incident light is assumed). **a–c** Extinction, scattering and absorption spectra of for different number of Au NPs. The *broken vertical line* is the resonant wavelength (555 nm) of single Au NP. **d** Normalized scattering intensity for respective number of Au NPs. **e** N-dependence on radiative width of a Au NP-array at the final position after a 0.1 s laser irradiation of 900 mW. The *second vertical axis* shows the enhancement factor in comparison with the case of single Au NP. Reprinted (adapted) with permission from [19]. Copyright (2012) American Chemical Society

be obtained for various polarization directions due to the high rotational symmetry. This result is convenient for solar energy conversion since the sunlight with broad spectrum and random polarizations can be efficiently collected. The obtained results and discussions provided in this subsection should pave the way toward a fabrication method of assembled structures of NPs whose optical properties correlate with various properties of a designed light field, i.e., "Light-induced-force material design". In the next subsection, we will introduce the experimental demonstration of our theoretical prediction.

5.3.3 Selective Optical Assembling of Uniform Nanoparticles

A natural photosynthetic bacterium has a LHA with a ring-like structure consisting of dye molecules [6, 7] whose absorption spectrum has changed through a selec-

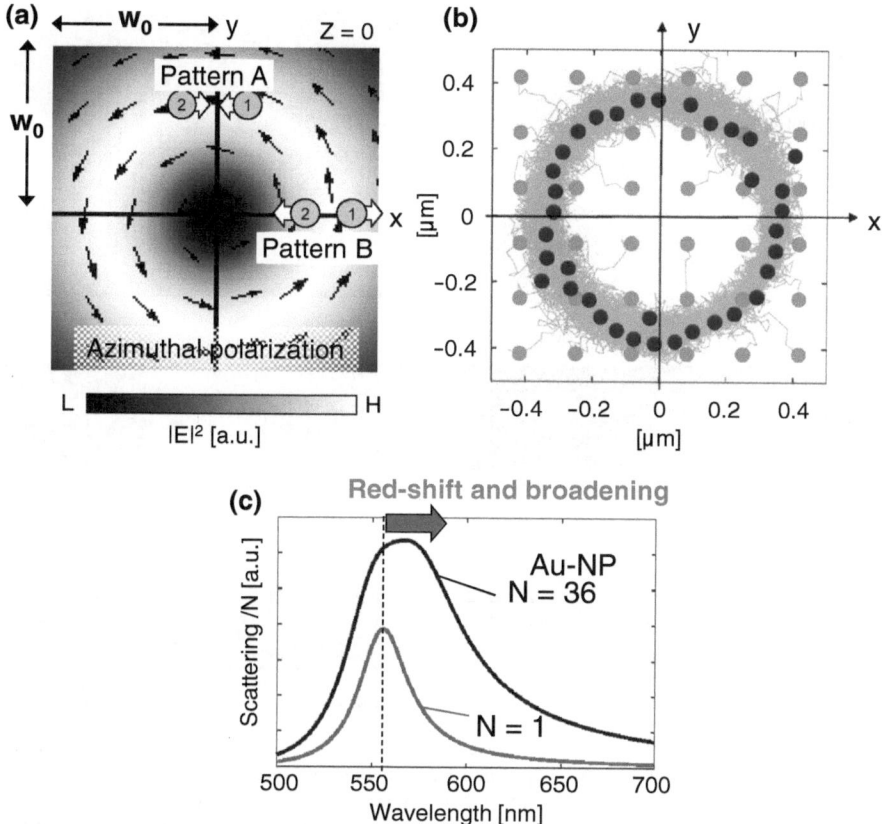

Fig. 5.12 **a** Spatial distribution of azimuthally polarized doughnut beam. **b** Ring-like configuration of Au NPs assembled by the doughnut beam of azimuthal polarization. **c** Normalized Scattering spectrum for $N = 36$ and that for $N = 1$

tive evolution [8] under the sunlight and thermal fluctuations as external stimuli depending on its territory. From this fact, we hit on an idea that we could selectively assemble non-biological nanomaterials exhibiting specific spectroscopic properties by tailored light and fluctuations. In this subsection, we will illustrate an experiment to demonstrate this idea for the selective assembling of ring-like arrangements of many anisotropic NPs with uniform shapes and orientations from a vast number of NPs with various shapes by using LIF with the help of thermal fluctuations.

The experimental configuration for the selective assembling of plasmonic nano-materials is shown in Fig. 5.13a. Ag nanostructures show a strong optical response with sharp LSP resonance due to a small nonradiative relaxation rate [38], and Ag NPs were prepared as target materials by a reduction method [55]. Before the experiment, it was confirmed by transmission electron microscope (TEM) that most of the prepared Ag NPs had shapes with low aspect ratios. A small number of nanos-tructures with high aspect ratios (>3) were also observed (Fig. 5.13b). The TEM

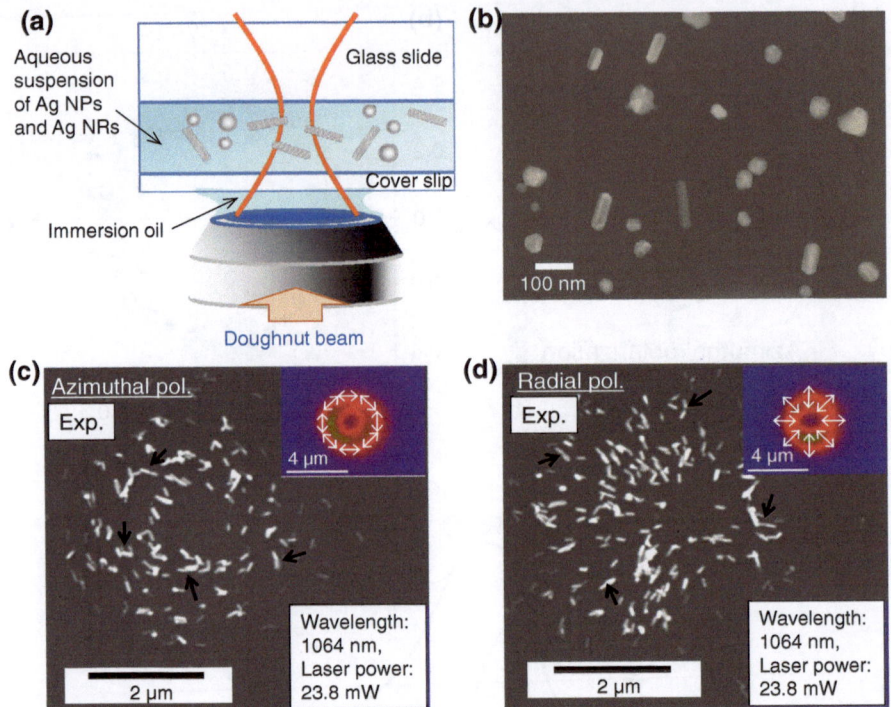

Fig. 5.13 a Schematic image of experimental setup for selective assembling of Ag NPs. Aqueous suspension of Ag NPs of various shapes and sizes, including Ag nanorods (NRs), are dropped between the glass slide and a cover slip. A Nd³⁺:YVO₄ laser (wavelength = 1064 nm, power = 23.8 mW) is used as the light source for optical tweezing and is focused on the substrate surface for the fixation of Ag NRs. An oil immersion lens is used as the objective lens for laser focusing. **b** Transmission electron microscope (TEM) image of Ag NPs used in this study without laser irradiation. **c** Scanning electron microscope (SEM) images of Ag NRs selected by the azimuthally polarized doughnut beam. (**d**) SEM images of Ag NRs selected by the radially polarized doughnut beam. The insets of (**c**) and (**d**) show the intensity distribution of the doughnut beam, as visualized by two photon imaging; and a schematic of the respective polarization

image shows various shapes existed, for example, spherical, triangular, and rod-like. However, after irradiation by the doughnut-like beam (azimuthal or radial polarization) for 6 minutes, nanostructures with specific characteristics were manipulated according to the properties of the beam. The majority of Ag NRs were concentrated in the region corresponding to the spatial distribution of the doughnut-like intensity distribution of diameter ∼2 μm. The remarkable point is that NRs of high aspect ratio (∼3) were fixed by the laser of 1,064 nm. Most interestingly, the majority of NRs were oriented to be parallel to the polarization direction of the azimuthal and radial (polarized) beams. Furthermore, many head-to-tail pairs of Ag NRs were observed, as indicated by the arrows in Fig. 5.13c, d.

In Fig. 5.14, an example of numerical simulations by LNMM in 5.2.2 showed that the LSP of a single Ag NR has a peak at 800 nm in water, and that when a

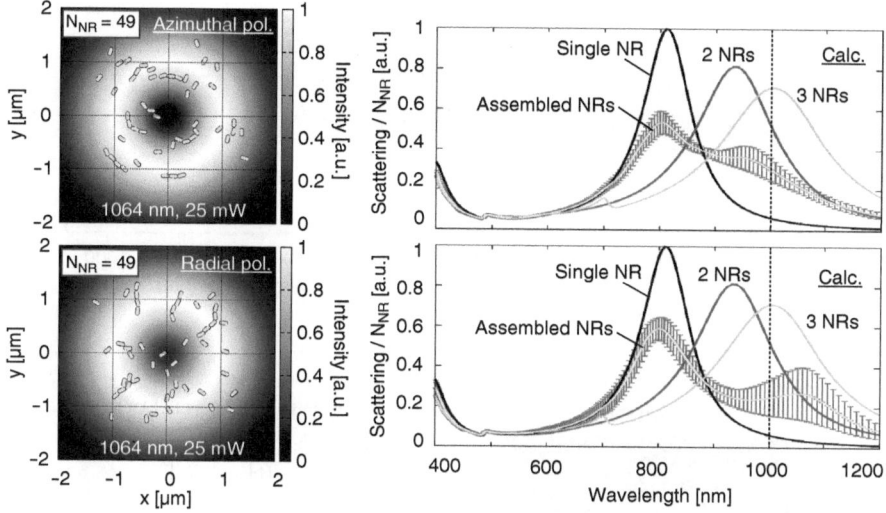

Fig. 5.14 (*Left*) Configurations of Ag NRs (40 nm wide, 120 nm long) by azimuthal and radial beams calculated by LNMM. (*Right*) Corresponding spectra of the total scattering from assembled Ag NRs. The number of NRs is NNR = 49. Also, spectra of a single Ag NR, and a head-to-tail array of two or three Ag NRs (with 30 nm separations) are shown together. In each line, the rotational average in xy-plane was taken under linear polarized light. *Vertical bars* in assembled NRs indicate standard deviation in rotational average

pair of Ag NRs were aligned in the coaxial direction and spaced 30 nm apart, the LSP resonance moved to about 920 nm. Also, the LSP resonance of three coupled Ag NRs is about 1,000 nm. Several Ag NRs are coupled with inter-object LIF and aligned parallel to the polarizations with a certain disorder arising from the thermal fluctuations as indicated by the arrows in Fig. 5.13c, d. Also, in order to understand the optical properties of the selectively assembled Ag NRs, their light scattering images and spectra were observed by dark-field optical microscopy (Fig. 5.15). The numerically calculated scattering spectra of assembled structure have similar shape as those in the experiment. The assembled Ag NRs with a ring-like shape appeared in optical images, as shown in the left figure of Fig. 5.15b. Corresponding scattering spectra show broad peak structures of LSP resonance from visible to infrared region (right figures in Fig. 5.15b). This behaviour is significantly different from the peak in the extinction of the original Ag NP suspension as shown in Fig. 5.15a owing to the selective deposition of NRs with high aspect ratios. As discussed in the explanation of Fig. 5.14 however, if multiple Ag NRs are closely spaced and fixed on the substrate it is expected that a large red-shift of the LSP resonance into infrared region can be observed. Also, a small peak appears at 700 nm is considered as an optical forbidden mode of three NRs from the calculation. These results imply that the multiple Ag NRs exhibit significantly strong electromagnetic coupling. Moreover, from Fig. 5.14, it can be seen that the Ag NRs assembled by the radially polarized beam exhibit a significantly red-shifted collective modes of LSPs resonant with the irradiated light

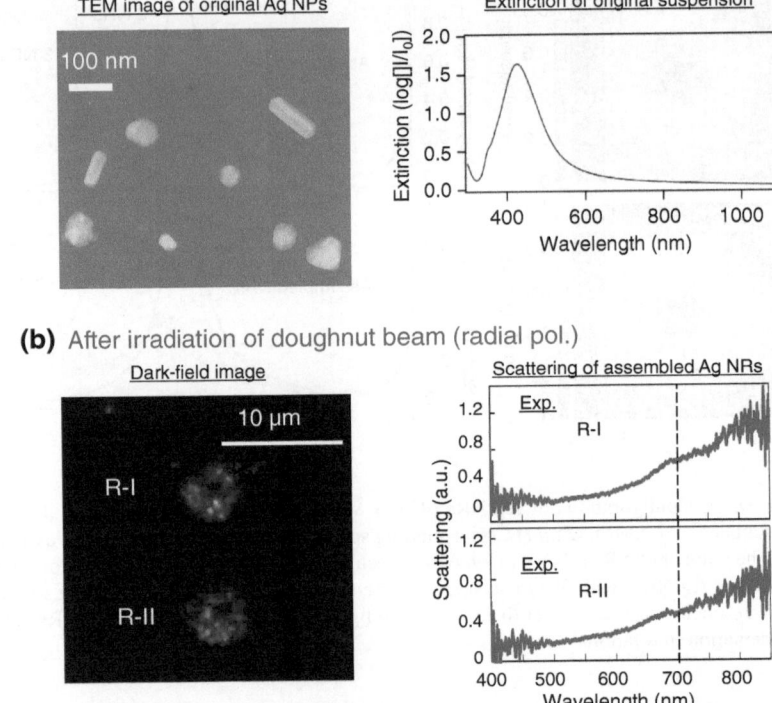

Fig. 5.15 **a** TEM image and Extinction spectrum of the suspension of Ag NPs. **b** *Dark-field optical image (scattering image)* and scattering spectra of ring-like arrangement of Ag NPs produced under the irradiation of radially polarized Nd^{3+}:YVO$_4$ laser with diameter $\sim$2 μm

and a spectral broadening due to the strong confinement in the radial direction is more prominent than the result of azimuthally polarized beam.

In order to understand the detailed mechanism of observed phenomena in Fig. 5.13, we show the theoretical evaluation of the spectral properties of LIF on Ag NRs of different aspect ratios (sphere, Ag NRs of aspect ratio 2 and 3) in Fig. 5.16. Two types of LIFs are shown here. One is the dissipative force arising from the photon momentum transfer in Fig. 5.16a, which pushes objects toward the laser propagation direction ($+z$-direction at {i}). The other is the gradient force (Fig. 5.16b) that is proportional to the light intensity gradient ($-x$-direction at {ii}). The gradient force attracts NRs toward the high intensity region leading to the trapping condition when the laser wavelength is longer than the resonance of the LSP corresponding to the peak of dissipative force. In the calculation, the long axis (length) and the short axis (width) of Ag NRs with high aspect ratio were set as 120 and 40 nm (in reference to Fig. 5.13b corresponding to NRs selected in Fig. 5.13c, d), whereas the length and width of Ag NRs with low aspect ratio were 80 and 40 nm. With experimentally used laser power, the optical potential well for Ag NRs of high aspect ratio parallel to the

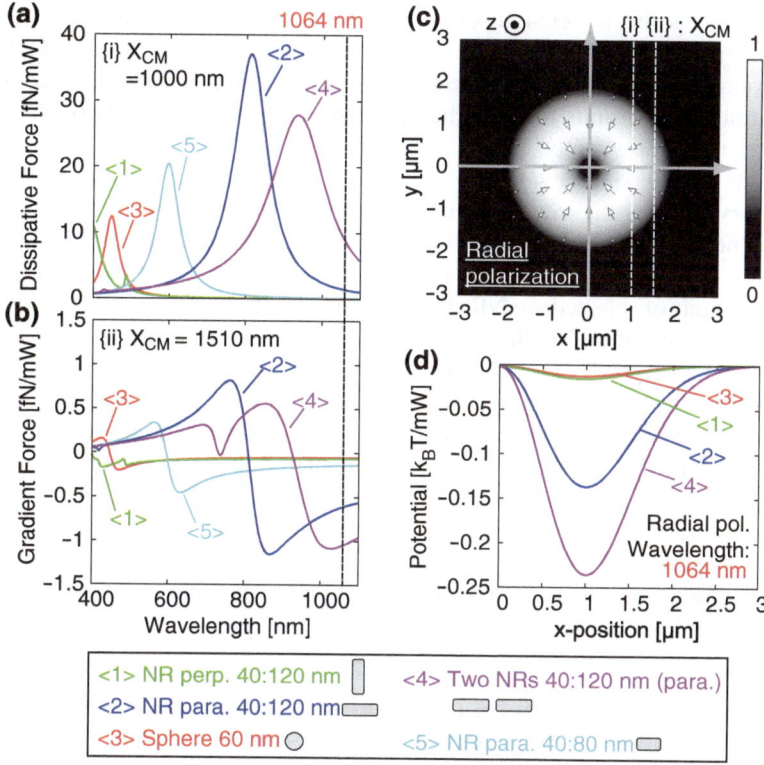

Fig. 5.16 Calculated spectra of the LIFs on Ag nanostructures. We consider a spherical NP (60 nm in diameter), a single NR with *low* aspect ratio (40 nm wide, 80 nm long) or with high aspect ratio (40 nm wide, 120 nm long) in azimuthally and radially polarized beams, respectively. A pair of Ag NRs of high aspect ratio with 30 nm separation is also considered. **a** Dissipative force on Ag nanostructures whose center of mass was located at {i}: ($X_{CM} = 1,000$ nm, $Y_{CM} = 0$, $Z_{CM} = 0$) in **c**, with the highest light intensity. **b** Gradient force on a spherical NP, a single NR, and two NRs whose center of mass was located at {ii}: ($X_{CM} = 1,510$ nm, $Y_{CM} = 0$, $Z_{CM} = 0$) in in **c**, with the steepest gradient of light intensity outside of the doughnut. **d** x-position dependence of optical potential by radial beam is plotted

light polarization is much deeper than the available thermal energy $k_B T$, (Fig. 5.16d; an indication of the strength of fluctuations in the suspension: $k_B T = 26$ meV, where the temperature $T = 298$ K). The optical potential wells for the spherical Ag NPs and the perpendicular Ag NRs are shallower, whereas the large trapping potential of the parallel Ag NRs results in a more effective selection process under the thermal fluctuations of $k_B T$. As shown in Fig. 5.16a, at 1,064 nm, the dissipative force on the parallel Ag NR of high aspect ratio (40:120 nm) is much greater than that on the spherical Ag NP, the perpendicular Ag NR, and the Ag NR of low aspect ratio (40:80 nm). Furthermore, when multiple Ag NRs are closely spaced and parallel to the polarization, the gradient force and the dissipative force for a pair of Ag NRs is greater than the case of a single NR due to a red-shift of the LSPs approaching

1064 nm that is the wavelength of Nd^{3+}:YVO_4. It is expected that the potential wells become deeper for more NRs aligned parallel to the polarization, and that the attractive and repulsive interparticle LIF as discussed in 5.2.1 had an effect on the orientation. These results indicate that a group of Ag NRs with a particular aspect ratio and orientation can be efficiently extracted by resonant LIF aided by thermal fluctuations. Also, similar selective assembling of Ag NRs with lower aspect ratio was observed under the irradiation of a single Gaussian beam with different wavelength of 660 nm [22]. Furthermore, the area of assembled structure of Ag NRs can be scaled down if more tightly focused beam. These results indicate that we can control the orientations of selected Ag NRs by changing the spatial distributions of polarization and intensity of an irradiated beam. These facts strongly support the selective arrangement of Ag NRs by LIF reflecting the tailored properties of manipulation light sources and leading to response optimized to them.

The experimental demonstration and theoretical analysis here will lead to a novel promising method for the selective assembling of various nanostructures with properties that are more optimized for the irradiated light. Such a selective assembling of uniform metallic nanostructures is desired in the wide field of nanotechnology, for example, the high performance light collecting system for solar energy conversion [56], the separation method of NPs useful for the medical applications [57], the biosensors and the catalysts [58].

5.4 Collective Phenomena in Metallic Nanoparticles Under Nonequilibrium Condition

5.4.1 Medical Applications

In the final section, we will discuss the biological applications of the collective phenomena of LSPs assembled metallic NPs. First, we introduce the medical application of the photothermal effect enhanced by interacting Au NPs in aqueous suspension [23, 25]. Recently, multi-nucleation of Au NPs at a very close distance can be chemically realized in PEG-attached dendrimer [59, 60]. Here, we will discuss the photothermal effect of such a closely spaced Au NPs and size dependence of generated heat and application for killing pathogenic cells.

Under the model with isolated cubic boxes containing N closely spaced multiple AuNPs (Fig. 5.17b), the total heat arising from the Au NPs can be described using the following equation:

$$H = \sum_{i=1}^{N} (R_{SV} N_{tot}/N) Q_i(\omega) L_S \propto \sum_{i=1}^{N} Q_i(\omega)/VN \qquad (5.22)$$

where $N_{tot} = MV_{SC}/\{(2\pi/3)(a_p/a_L)^3\} \propto V^{-1}$ is the total number of NPs [61] assuming the same diameter $d_i = a_p$ and $V_i = V$ for all the NPs, M is the concentration of Au atoms (110 μM), V_{SC} is the volume of solvent in the sample cell (3 mL),

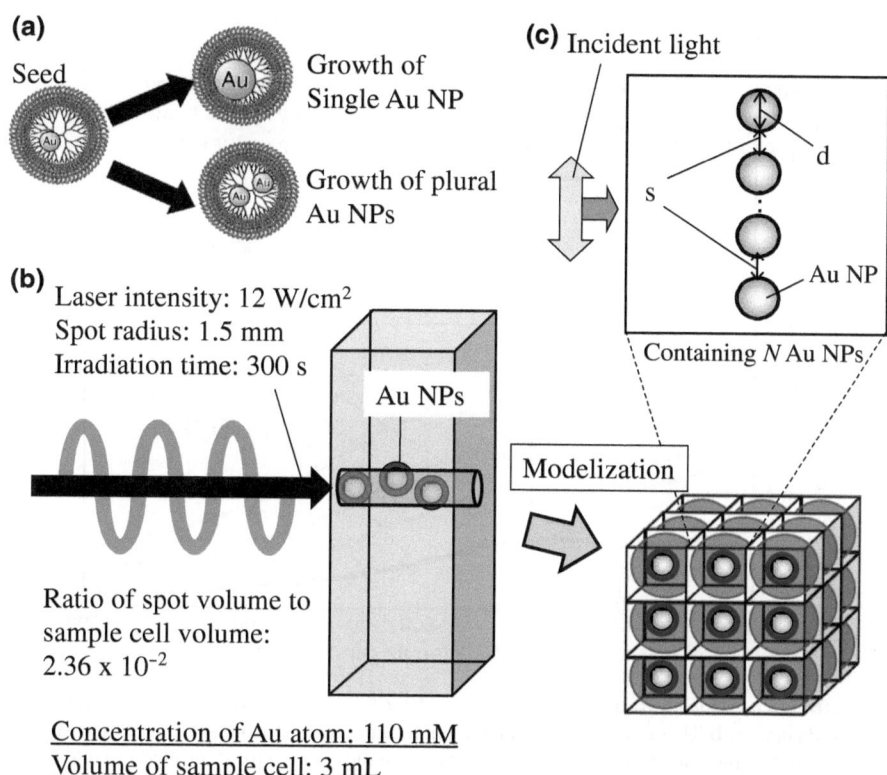

Fig. 5.17 **a** Schematic image of seeding growth in the PEG-attached dendrimer. **b** Schematic image of the model of the photothermal effect. The modeled system for the calculation is also shown. **c** Model for multiple Au NPs. Reprinted (adapted) with permission from [23]. Copyright (2011) American Chemical Society

a_L is the edge length of a unit cell of crystal of Au, R_{SV} is the rate of laser spot volume to sample cell volume (2.36×10^{-2}), and L_S is the laser irradiation time (300 s). The background dielectric function $\varepsilon_{bg}^{(p)}(\omega)$ in (5.3) was estimated from the experimental values with interband effects that were observed in bulk Au [38, 39]. Other optical parameters are the same as those used in previous subsections. $R_{SV} N_{tot}/N$ indicates the number of cubic boxes in the laser spot and $\sum_{i=1}^{N} Q_i(\omega)L_S$ indicates the total heat in each cubic box. A laser intensity is assumed to be 12 W/cm² as shown in Fig. 5.17b. We can evaluate the heat per unit volume from the equation of Joule heat [62] as follows

$$Q_i(\omega)/V = (\omega/2)\mathrm{Im}[\chi_i] \cdot |\mathbf{E}_i|^2. \tag{5.23}$$

Next, in Fig. 5.18, the photothermal effect from multiple Au NPs was numerically investigated at different distances by considering Green's function in (5.4) for a small value of $|\mathbf{r}_{ij}|$. Given that there are two neighboring Au NPs with diameters of 2 nm at

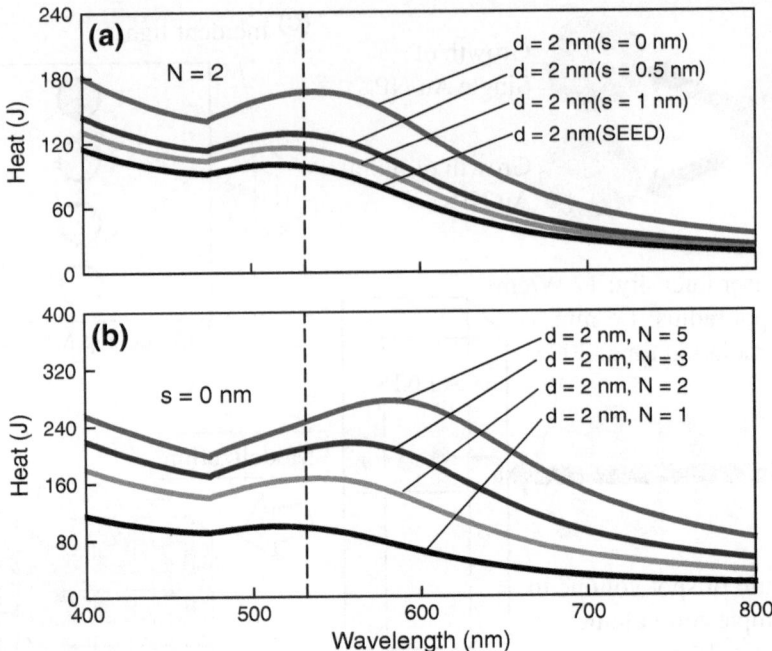

Fig. 5.18 a Theoretical results for photo-induced heat from pairs of Au NPs for different inter-particle distances. **b** Theoretical results for the photo-induced heat from multinucleated Au NPs for respective values of N. The *vertical dashed lines* indicate 532 nm. Reprinted (adapted) with permission from [23]. Copyright (2011) American Chemical Society

different distances (here, we set the distance between the nearest neighbor NPs to be $|\mathbf{r}_{i,i+1}| = s$), the photo-thermogenic properties were affected by the distance. The red-shift and the peak value in heat spectra increases as the decrease of interparticle distance s, and such a red-shift arises from energy shift due to the attractive interaction via $\mathbf{G}^{\mathrm{med}}(s)$ among the induced polarizations of LSP in multiple Au NPs. In addition, spectral broadening arises from $2V_{\mathrm{f}}/a_{\mathrm{p}}$ with a large value for small particle diameter, and $\bar{\Gamma}_{\mathrm{int}}$ has a large value for small s in (5.8). Such a tendency indicates that the enhancement of heat in near-infrared region, as the interparticle distance decreases, arises from the difference of background dielectric function of a Au NP and that of surrounding medium providing multiple interactions of LSPs between Au NPs due to the effect of mirror images at the surface of Au NPs. This means that the heat effect can be controlled by the design of surface protecting agent or environmental refractive index. Furthermore, considering the model in Fig. 5.17, the photo-induced heat generation of multiple contacting Au NPs with diameters of 2 nm for different number of Au NPs was investigated. A very large red-shift and enhancement of the photo-induced heating effect were obtained by increasing the number of Au NPs in each cubic box since the interaction of mirror images of localized surface plasmon is greatly enhanced. These results indicate that closely spaced multiple small Au NPs would be useful since they exhibit large red-shift and efficient discharge from

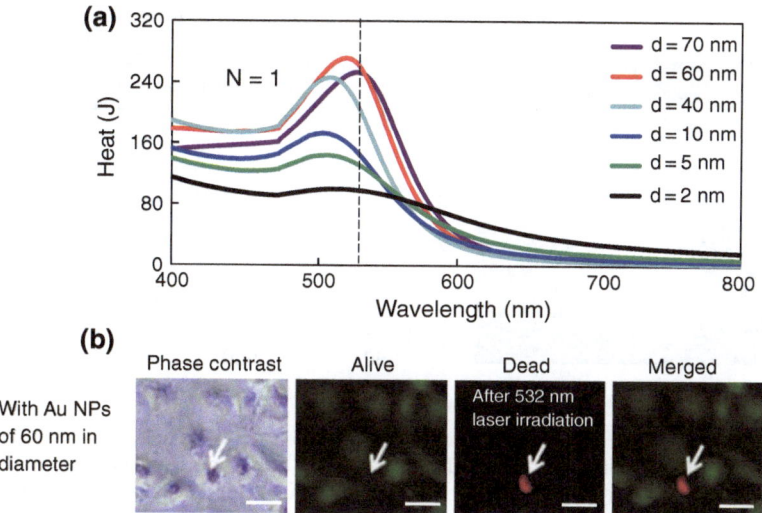

Fig. 5.19 **a** Calculated results for the photo-induced heat from single Au NPs with different diameters. The *vertical dashed line* indicates 532 nm. Reprinted (adapted) with permission from [23]. Copyright (2011) American Chemical Society. **b** Experimental results of observation of dead or alive in HeLa cell with Au NPs of 60 nm in diameter under the irradiation of 532 nm laser light. The scale bars indicate 30 μm

a human body. Since the enhancement of the heat effect in near-infrared region is indispensable for in vivo applications, larger red-shift is expected by controlling the interaction of LSPs in multiple metallic NPs.

On the other hand, paying attention to the photothermal effect of single Au NPs (Fig. 5.19a), the generated heat was largely improved by controlling the size of the Au NPs near 60 nm [23]. We performed preliminary experiment of photothermal therapy by using the commercial Au NPs [25]. Under the assumed condition, the Au NP with a 60 nm diameter (48 nm in TEM analysis) exhibited the highest photo-thermogenic properties by 532 nm laser irradiation according to our investigations. Thus, there is an optimum size for the most efficient photo-thermogenic properties. In fact, we have confirmed that HeLa cell can be destroyed by doping similar sized Au NPs under 532 nm laser irradiation (Fig. 5.19b). While these results are preliminary, they provides important guidelines for the design of suitable Au NPs for photothermal therapy and various types of biomedical applications.

5.4.2 Biosensor Applications

In the final part of this chapter, we will briefly explain the novel mechanism of highly-sensitive optical biosensors based on the coupling of heterogeneous metallic nanostructures, i.e., Au nanorods (NRs) and Ag NP-fixed beads (AgNP-FB) [26], and the light-induced bubble to control the local phase transition by using photothermal

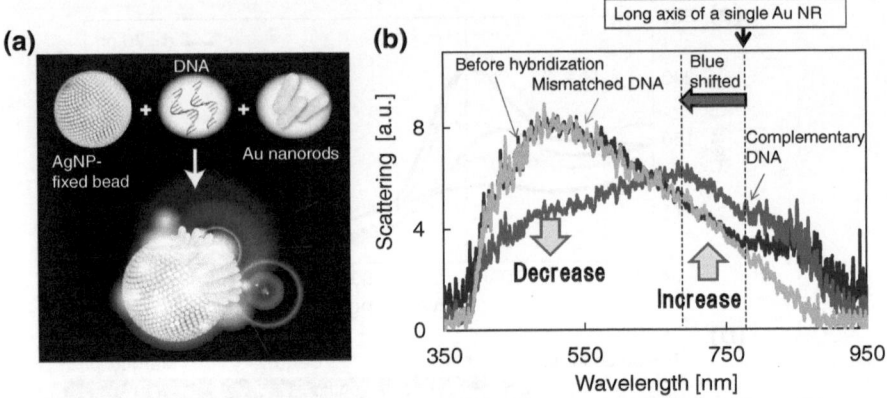

Fig. 5.20 a Schematic image of coupling of Au nanorod (AuNR) and Ag nanoparticle-fixed bead (AgNP-FB) with hybridization of thiolated-DNA. **b** Scattering spectra of AgNP-FB before hybridization and after hybridization. The result obtained using the mismatched DNA is also shown. Reprinted (adapted) with permission from [26]. Copyright (2014) American Chemical Society

effect of AgNP-FB [27] as an example of three-dimensional biomimetic optical manipulation.

Figure 5.20a schematically illustrates the mechanism by which AuNRs bind to the surface of AgNP-FB, where the probe DNA (A) $5'$-SH-poly(T)$_{12}$-$3'$ and the probe DNA (B) $3'$-SH-poly(T)$_{12}$-$5'$ were adsorbed onto the both transverse surfaces of a single AuNR (30×10 nm) and the surface of AgNP-FB (mean diameter of Ag NPs: $\sim$5 nm, mean diameter of core plastic bead: 5 μm), respectively. The complementary DNA (24mer $5'$-poly(A)$_{24}$-$3'$) and mismatched DNA (24mer $5'$-poly(T)$_{24}$-$3'$) were used as the target. For clarification of the spectral modulation of LSP and spatial configurations under heterogeneous coupling of AgNP-FB and AuNRs in aqueous solution, we performed dark-field optical spectroscopy to observe the light scattering from a single AgNP-FB using different kinds of target DNA (Fig. 5.20b). Before hybridization, the main AgNP-FB peak appears around 490 nm in the scattering spectrum, there is almost no change after addition of the mismatched DNA despite a small modulation of the long wavelength region. On the other hand, we observed a clear modulation in wavelength after adding complementary DNA. Remarkably, the main AgNP-FB peak was greatly suppressed whereas another peak near the LSP resonance of the long-axis mode of AuNR (around 750 nm) was significantly enhanced. Also, the change in the surface conditions of a single AgNP-FB was confirmed by using a scanning electron microscope (SEM) before and after the addition of target DNA into the mixture of AuNRs and AgNP-FB whose surfaces were modified with DNA probes. Before addition of DNA, Ag NPs were uniformly assembled on the surface of plastic beads, and there is no change when mismatched DNA was used. However, after addition of complementary DNA, many anisotropic NPs were observed on the AgNP-FB in the SEM image and a clear signal for Au appeared in the EDX measurement [26]. These results indicate that AuNRs with

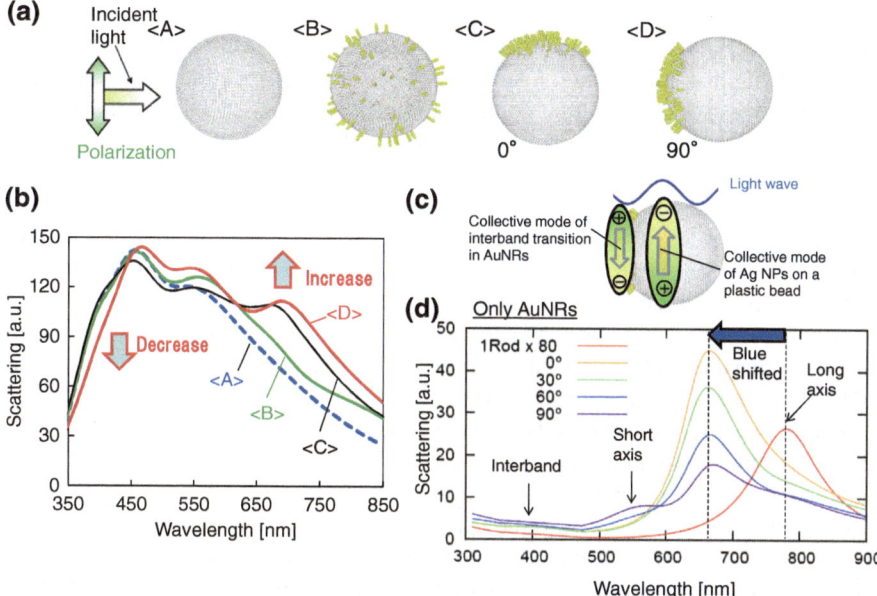

Fig. 5.21 **a** Models for the calculations. **b** Calculated scattering spectra of Ag NP-fixed beads (AgNP-FB) without Au NRs <A> and with standing Au NRs the geometry <B>–<D>, where 1 nm separation between a AgNP and a AuNR was assumed to be similar in size to the DNA. **c** Schematic illustration of the cancellation of collective modes of interband transition and long-axis LSP. **d** Calculated scattering spectra of only Au NRs without AgNP-FB. Reprinted (adapted) with permission from [26]. Copyright (2014) American Chemical Society

probe DNA (A) were conjugated onto the AgNP-FB with probe DNA (B) due to their hybridization with complementary DNA.

Furthermore, in order to understand the physical mechanism behind this unexpected spectral modulation, we systematically investigated the light scattering from the coupled system of AuNRs and AgNP-FB based on the Cluster-DISC method as explained in 5.2.3 (Fig. 5.21). We investigated the scattered light from a coupled system of AgNP-FB (diameter of AgNPs: 2.5 nm, diameter of core bead: 400 nm) and AuNRs with various spatial configurations <A>–<D> in Fig. 5.21a. In order to discuss the essential mechanism by which this occurs and the theoretical limitation for the detection of the modulation, we assumed a model system whose size was smaller than the experimentally observed sample consisting of a coupled system of AgNP-FB and AuNRs. A similar treatment successfully explained the collective phenomena of LSP in a AuNP-fixed bead [24]. It was confirmed that the calculated optical spectrum of AgNP-FB without AuNRs as shown in <A> well corresponds to the experimental results using AgNP-FB with a diameter of 400 nm. In particular, as shown in the configuration of <D>, AuNRs are conjugated on the surface at the side of incident light source. In Fig. 5.21b, we found that the spectrum in the short wavelength region between 350 and 400 nm decreases and that a peak at longer wavelength region near the LSP resonance of AuNRs (approximately 750 nm) appears. This tendency is similar to the experimental result in Fig. 5.20b since the incident light was irradiated from the bottom of the glass slide in the dark-field measurement. Also, in a different configuration as shown in <C>, the enhancement in the longer wavelength region appears and the decrease in the short wavelength region appears between 400 and 500 nm although the scattering increases over the wide wavelength region, and the decrease in the short wavelength region is quite negligible in the low density case shown in <B>. Therefore, we consider that AuNRs on a AgNP-FB at the configuration <C> also affected on the decrease of the peak in short wavelength region. If lying AuNRs are assumed, the enhancement of scattering appeared over the considered wavelength region, which was significantly different from the experimentally obtained spectrum. Therefore, this tendency implies that AuNRs remain upright at the surface of AgNP-FB.

In order to analyze the detailed mechanism, we assumed a model in which AuNRs were in the same configuration as that seen in Fig. 5.21b, whereas the AgNP-FB were neglected (Fig. 5.21d). For the high density case with angles 0–90°, the peak position between 700 and 800 nm shifts to the wavelength region shorter than 700 nm due to the repulsive interaction since the standing rods have components of induced polarization of LSP that are parallel to the incident light polarization. In the case of the high angle, the rate of AuNRs perpendicular to light polarization increases. Therefore, the peak of short-axis mode increases and a peak of the long-axis mode decreases. In addition, the interband component gets larger by increasing the angle from 0 to 90°. The result without interaction between AuNRs is also shown, whereas the scattering in the case of densely assembled AuNRs is much larger. Taking into account these calculated results, the decrease of scattering in the short wavelength region can be attributed to the cancelation of the collective mode of interband absorption and the collective modes of LSP in the AgNP-FB as shown in the schematic image in Fig. 5.21c. While the experimentally observed suppression is more prominent than the calculated value, it is considered that such a difference appears due to the underestimation of the interband effect in the simulation using the parameters obtained from the bulk values in [38]. The quenching of the dye molecules has been reported under certain conditions involving coupling with Au NPs [63], but our findings may be attributed to a completely different mechanism. On the other hand, the enhancement of peak in long wavelength regions arises from the plasmonic superradiance from long-axis mode in AuNRs, which is similar to the high density Au NPs [19, 24] and Ag NRs [22].

Furthermore, we explored the possibility of application for an optical biosensors. From Fig. 5.21b, detectable spectral modulations arose even when 80 Au NRs were conjugated to AgNP-FB using a similar amount of DNA. The number of DNA molecules attached to the edge of the AuNR was estimated to be 1–8 from the area density of several tens pmol/cm^2 of an experimentally used sample [64]. This result demonstrates that zmol ($\sim 6 \times 10^2$) levels of DNA can be detected by using a scattering spectrum under a white light irradiation and provides a guiding principle for a highly sensitive biosensor. This demonstrates the feasibility of highly sensitive detection of DNA on the order of zmol with a simple optical system using white light. The

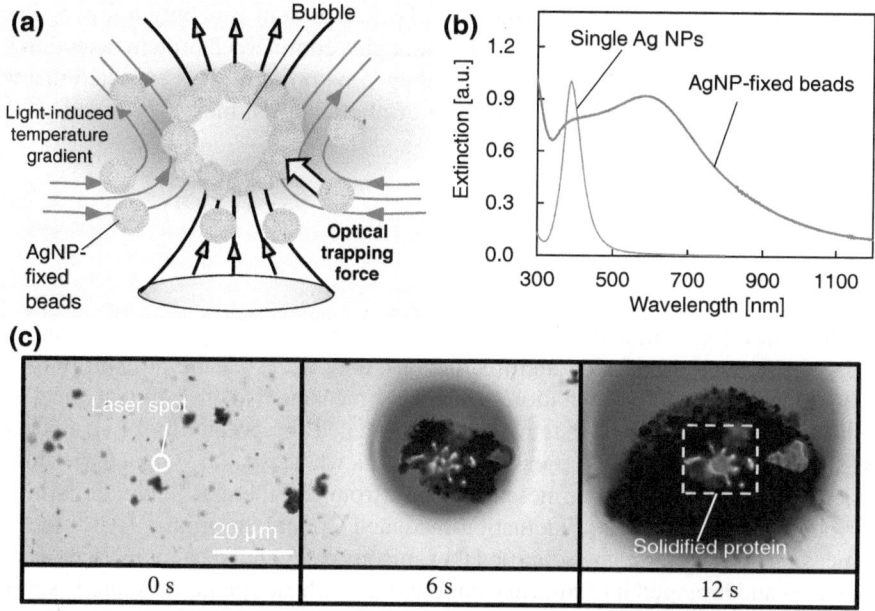

Fig. 5.22 **a** Schematic of control of photothermal phase-transition. **b** Extinction spectra of AgNP-FB and single Ag NPs in aqueous suspension. **c** Snapshots of dynamics of AgNP-FBs in the aqueous solution of albumin under the infrared laser irradiation. Reprinted (adapted) with permission from [27]. Copyright (2014) American Chemical Society

results and discussion in this contribution will pioneer a convenient and sensitive optical biosensing system based on the biomolecule-mediated self-assembly of heterogeneous metallic nanostructures.

Meanwhile, paying attention to photothermal effect of densely assembled metallic NPs as discussed in 5.4.1, where the absorption spectrum of LSP shows great broadening and red-shift even in AgNP-FB (Fig. 5.22b). Utilizing such properties of AgNP-FB, we have clarified the photothermal assembling phenomena and submillimeter three-dimensional bubble generation [27] under the irradiation of infrared light. Since such a bubble structure is similar to a vesicle in biological cells, there is a possibility for containing biological materials therein. Figure 5.22a shows the schematic illustration of such an assembly process. A gradual convection flow was caused by the laser-induced heat just after irradiation, and this flow transported AgNP-FBs to the region near optical trap by LIF at the focal point of the laser. The photothermal effect of the optically-trapped AgNP-FBs accelerates the convection flow and leads to the generation of a macroscopic bubble. The bubble size gradually increased but convection did not stop due to the enhancement of the photothermal effect as a result of the assembled AgNP-FBs. As a result, in Fig. 5.22c, we have successfully demonstrated the photothermal solidification of proteins (albumin) whose solidification point is about 78 °C as analytes by high plasmonic photothermal effect. The roughly estimated weight of solidified protein was only a few pg even within 1 min by our developed principle although a commercial protein detection kit takes several hours for the detection of 100 pg–5 ng proteins. It is expected that the obtained results will pioneer a novel type of photothermal biosensor based on the three-dimensional optical bottom-up process.

5.5 Summary

We have developed new theoretical method to describe the dynamics of nanomaterials under light irradiation and fluctuations, and discussed the possibility of the "Biomimetic Optical Manipulation" from one-dimensional to three-dimensional systems. In one dimensional system, we have revealed the possibility of size-sorting NPs by designing multiple optical potential wells whose spatial symmetry is slowly modulated in the diffusion timescale arising from the thermal fluctuations in the surrounding medium, i.e., "Fluctuation-mediated Optical Screening". Also, in two-dimensional system, we have clarified that anisotropic metallic nanomaterials can be extracted and arranged into ring-like configuration whose spatial structure is similar to a natural light harvesting antenna and exhibit significantly large red-shift with spectral broadening optimum for the irradiated light. This principle is useful for the creation of efficient light-energy conversion materials and biological applications. In addition, we have discussed about the biomedical and the biosensing applications of collective phenomena of localized surface plasmons for the enhancement of the photothermal effect to kill the pathogenic cells in the drug delivery system, and for the enhancement of their spectral modulation to detect DNA or small biological molecules. Furthermore, very high photothermal effect of densely assembled metallic NPs can be used for the generation of macroscopic bubble and the solidification of proteins as examples of laser-induced nonequilibrium systems, which can be considered as the example of three-dimensional biomimetic optical manipulation. The results and discussion here will open new avenues for an interdisciplinary research field of the nonbiological nanophotonics and the biological nanoscience based on the light-induced nonequilibrium dynamics.

Acknowledgments The authors would like to thank their students, Mr. M. Tamura, Mr. Y. Nishimura, Mr. H. Hidaka, Mr. H. Hattori, Dr. H. Yamauchi, Mr. K. Nishida, Mr. T. Hamada, Mr. N. Oeda, and Mr. Y. Watanabe for their help in the research summarized here. Also, they would like to thank Prof. T. Tsutsui, Prof. H. Ishihara, Dr. T. Itoh, Prof. H. Miyasaka, Dr. Y. Yamamoto, Prof. Shiigi and Prof. T. Nagaoka for kind encouragement and support. This work was supported by PRESTO from the JST; a Grant-in-Aid for Scientific Research (B) No. 23310079 and No. 26286029; a Grant-in-Aid for Young Scientists (A) No. 23681023; the Grants-in-Aid for Exploratory Research No. 23655072 and No. 24654091 and No. 26610089 from the JSPS; and Special Coordination Funds for Promoting Science and Technology from the MEXT (Improvement of Research Environment for Young Researchers (FY 2008–2012).

References

1. G. Nicolis, I. Prigogine, *Self-Organization in Non-Equilibrium Systems* (Wiley, New York, 1977)
2. E. Rabani, D.R. Reichman, P.L. Geissler, L.E. Brus, Nature **426**, 271 (2003)
3. E.V. Shevchenko, D.V. Talapin, N.A. Kotov, S. O'Brien, C.B. Murray, Nature **439**, 55 (2006)
4. P.W.K. Rothemund, Nature **440**, 297 (2006)
5. N. Kodera, D. Yamamoto, R. Ishikawa, T. Ando, Nature **468**, 72 (2010)
6. S. Bahatyrova et al., Nature **430**, 1058 (2004)
7. S. Scheuring, J.N. Sturgis, Science **309**, 484 (2005)
8. N.Y. Kiang, J. Siefert, Govindjee, R. E. Blankenship. Astrobiology **7**, 222–251 (2007)
9. M. Rosoff (Ed.), *Vecicles* (Marcel Dekker, Inc., New York, 1996)
10. A. Ashkin, Phys. Rev. Lett. **24**, 156 (1970)
11. S. Ito, H. Yoshikawa, H. Masuhara, Appl. Phys. Lett. **80**, 482 (2002)
12. T. Iida, H. Ishihara, Phys. Rev. Lett. **90**(057403), 1–4 (2003)
13. K. Inaba et al., Phys. Stat. Sol. (b) **243**, 3829–3833 (2006)
14. T. Iida, H. Ishihara, Phys. Rev. Lett. **97**, 117402 (2006). Phys. Rev. B **77**, 245319 (2008)
15. S. Ito, Y. Tanaka, H. Yoshikawa, Y. Ishibashi, H. Miyasaka, H. Masuhara, J. Am. Chem. Soc. **133**, 14472 (2011)
16. D.G. Grier, Nature **424**, 810–816 (2003)
17. Y. Jiang, T. Narushima, H. Okamoto, Nature Phys. **6**, 1005 (2010)
18. Y. Tsuboi et al., J. Phys. Chem. Lett. **1**, 2327 (2010)
19. T. Iida, J. Phys. Chem. Lett. **3**, 332 (2012)
20. M. Tamura, S. Ito, S. Tokonami, T. Iida, Res. Chem. Intermed. **40**, 2303 (2014)
21. M. Tamura, T. Iida, Nano Lett. **12**, 5337 (2012)
22. S. Ito, H. Yamauchi, M. Tamura, S. Hidaka, H. Hattori, T. Hamada, K. Nishida, S. Tokonami, T. Itoh, H. Miyasaka, T. Iida, Sci. Rep. **3**, 3047 (2013)
23. C. Kojima, Y. Watanabe, H. Hattori, T. Iida, J. Phys. Chem. C **115**, 19091 (2011)
24. S. Tokonami, T. Iida et al., J. Phys. Chem. C **117**, 15247 (2013)
25. C. Kojima, N. Oeda, S. Ito, H. Miyasaka, T. Iida, Chem. Lett. **43**, 975 (2014)
26. S. Tokonami, K. Nishida, S. Hidaka, Y. Yamamoto, H. Nakao, T. Iida, J. Phys. Chem. C **118**, 7235 (2014)
27. Y. Nishimura, K. Nishida, Y. Yamamoto, S. Ito, S. Tokonami, T. Iida, J. Phys. Chem. C **118**, 18799 (2014)
28. D. Ermak, J. Chem. Phys. **69**, 1352 (1978)
29. J.N. Israelachvili, *Intermolecular And Surface Forces*, 2nd edn. (Academic Press, London, 1992)
30. T. Kim, K. Lee, M. Gong, S.W. Joo, Langmuir **21**, 9524 (2005)
31. A.J. Hallock, P.L. Redmond, L.E. Brus, Proc. Natl. Acad. Sci. USA **102**, 1280 (2005)
32. L. Novotny, C. Henkel, Opt. Lett. **33**, 1029 (2008)
33. Z. Li, M. Kall, H. Xu, Phys. Rev. B **77**, 085412 (2008)
34. E.M. Purcell, C.R. Pennypacker, Astrophy. J. **186**, 705 (1973)
35. J.J. Goodman, B.T. Draine, P.J. Flatau, Opt. Lett. **16**, 1198 (1991)
36. N.B. Piller, O.J.F. Martin, Phys. Rev. E **58**, 3909 (1998)
37. T. Iida, Y. Aiba, H. Ishihara, Appl. Phys. Lett. **98**, 053108 (2011)
38. P.B. Johnson, R.W. Christy, Phys. Rev. B **6**, 4370 (1972)
39. R. Antoine, P.F. Brevet, H.H. Girault, D. Bethell, D.J. Schiffrin, Chem. Commun. **1997**, 1901 (1997)
40. K. Svoboda, S.M. Block, Opt. Lett. **19**, 930 (1994)
41. Y. Yamamoto, S. Takeda, H. Shiigi, T. Nagaoka, J. Elect. Soc. **154**, D462 (2007)
42. G.F. Bohren, D.R. Huffman, *Absorption and Scattering of Light by Small Particles* (Wiley Interscience, New York, 1983)
43. L. Gammaitoni, P. Hanggi, P. Jung, F. Marchesoni, Rev. Mod. Phys. **70**, 223 (1998)

44. H. Yoshikawa, T. Matsui, H. Masuhara, Phys. Rev. E **70**, 061406 (2004)
45. Y. Seol, A.E. Carpenter, T.T. Perkins, Opt. Lett. **31**, 2429 (2006)
46. T. Uwada, T. Sugiyama, H. Masuhara, J. Photochem. Photobiol. A: Chem. **221**, 187 (2011)
47. R.H. Dicke, Phys. Rev. **93**, 99 (1954)
48. Y. Ohfuti, K. Cho, Phys. Rev. B **51**, 14379 (1995)
49. M. Scheibner, T. Schmidt, L. Worschech, A. Forchel, G. Bacher, T. Passow, D. Hommel, D. Nat, Phys. **3**, 106 (2007)
50. K. Miyajima, Y. Kagotani, S. Saito, M. Ashida, T. Itoh, J. Phys. Condens. Matter. **21**, 195802 (2009)
51. C. Dahmen, B. Schmidt, G. von Plessen, Nano Lett. **7**, 318 (2007)
52. R. Oron et al., Appl. Phys. Lett. **77**, 3322 (2000)
53. B. Gu, Y. Cui, Opt. Exp. **20**, 17684 (2012)
54. H. Hattori, S. Hidaka, T. Iida, IQEC/CLEO Pacific Rim 2011 Tech. Dig. **2011**, 1817 (2011)
55. P.C. Lee, D. Meisel, J. Phys. Chem. **86**, 3391 (1982)
56. H.A. Atwater, A. Polman, Nat. Mat. **9**, 205 (2010)
57. K. Weintraub, Nature **495**, S14 (2013)
58. S. Tokonami, Y. Yamamoto, H. Shiigi, T. Nagaoka, Anal. Chim. Acta **716**, 76 (2012)
59. Y. Haba, C. Kojima, A. Harada, T. Ura, H. Horinaka, K. Kono, Langmuir **23**, 5243 (2007)
60. Y. Umeda, C. Kojima, A. Harada, H. Horinaka, K. Kono, Bioconjug. Chem. **21**, 1559 (2010)
61. B.D. Chithrani, A.A. Ghazani, W.C.W. Chan, Nano Lett. **1**, 84 (2006)
62. J.D. Jackson, *Classical Electrodynamics*, 3rd edn. (Wiley, New York, 1999)
63. S. Mayilo et al., Nano Lett. **9**, 4558 (2009)
64. S. Tokonami, H. Shiigi, T. Nagaoka, Anal. Chem. **80**, 8071 (2008)

Index

© Springer International Publishing Switzerland 2015
M. Ohtsu and T. Yatsui (eds.), *Progress in Nanophotonics 3*,
Nano-Optics and Nanophotonics, DOI 10.1007/978-3-319-11602-0